AUDIT PROTOCOL FOR THE INVESTIGATION OF CONTAMINATED SITES

A Case Study

Dr. Ir. Amar Singh Toor

ISBN
978-1-5437-5067-6 (sc)
978-1-5437-5066-9 (e)

Print information available on the last page.

To order additional copies of this book, contact
Toll Free 800 101 2657 (Singapore)
Toll Free 1 800 81 7340 (Malaysia)
www.partridgepublishing.com/singapore
orders.singapore@partridgepublishing.com

05/21/2019

PARTRIDGE

DEDICATION

This book is dedicated to my late sisters Chani and Mindy,
who had sponsored my engineering degree course at Middlesex U, London.
We treasure your love and
you will always be part of my life,
till I depart.

PREFACE

I would like to thank my mentors, peers, colleagues and associates in encouraging me in my pursuit to write this book. Special thanks to my dearest associates, Professor Ir. Dr. Omar Kadir and Professor Dr. Nik Norulaini, without your encouragement this would not be possible.

Professor Omar was always there to listen to my provoking thoughts and ideas, provided guidance and advice. Without his support this work would have been impossible to accomplish. Professor Dr. Nik's encouragement kept me focussed on my pursuit.

I thank my family (Neeta, Kelvin, Meenu, Anil, Neetaraj, Sunil and Shareen) for the unconditional support, encouragement to pursue my interest and believing in my pursuit. The times I kept to myself working through the days and nights away from all of you – in preparation to fulfil my dreams.

My little princess Mysha, with you around me, made my day joyous. I look forward for each sunrise to be with you.

My dearest Kiren, you had always been a great source of encouragement and an example setter for the family. Thank you.

My siblings Joe, Gurdip and Gudi for being with me and spending memorial times with me. I treasure those moments of my life.

Biography

Dr. Ir. Amar is a leader with more than 35 years of work experience. Extensive global experience in more than 10 countries, leading cross-functional teams and industry-wide initiatives to develop product integrity and EHS sustainability strategies, business continuity plans, lean manufacturing, value stream mapping and social compliance programmes.

Dr. Ir. Amar is a recognised expert in industrial engineering, product development and product integrity, EHS, social compliance and lean manufacturing. He has a strong balance of technical credibility and experience to drive progress on sustainability strategies and challenges that multi-national corporations face today. He is an excellent applicator and trainer on problem solving methodologies and problem solving tool applications. Process demonstrated abilities to create and communicate innovative strategies across the supply chains of globally recognised brands. Persuasive advocate and collaborative partner focussed on aligning policy initiatives with business objectives in order to maximise benefit to the enterprise and its stakeholders.

Dr. Ir. Amar currently leads ROOT Business Solutions (RBS) as their principal consultant by providing expertise on subject matters in relation to product development and compliance, manufacturing, engineering, value stream mapping and resolving/eliminating business or process related issues. Prior to this assignment, he was attached to Mattel Inc as a Senior Director. He led the South East Asia Product Integrity and Asia EHS & Compliance teams for the world's largest toy company. Played an instrumental role in planning, establishing requirements and developing protocols to be used by Mattel own operated plants, vendors and subcontractors. He collaborated with functional leaders and stake holders across regions to ensure compliance to local regulations, local practices and Mattel requirements. He represented Mattel at government and standards ministry departments level in Malaysia, Singapore, India and Indonesia. He worked with the respective standards development departments to establish local toy standards.

Dr. Ir. Amar obtained his doctorate in environmental technology from USM, Penang. He has a master of Science degree in quality control and instrumentation. He obtained his Bachelor's degree in Engineering from Middlesex U, London.

Dr. Ir. Amar is a registered professional engineer in Malaysia and was a registered chartered professional engineer in Australia.

Dr. Ir. Amar Singh Ph. D, PE, MIEM – ROOT Business Solutions

Email: amarsinghtoor@gmail.com

CONTENTS

Appendixes

LIST OF ABBREVIATIONS

ACM	Asbestos Contaminated Material
AHERA	Asbestos Hazard Emergency Response Act
ARAR	Applicable or Relevant and Appropriate Requirements
ASTM	American Society for Testing and Materials
bgl	below ground level
CERCLA	Comprehensive Environment Response, Compensation and Liability Act (Superfund)
CI	Confidence Interval
CLARINET	Contaminated Land Rehabilitation Network for Environmental Technologies
CLEA	Contaminated Land Exposure Assessment
COC	Chain of Custody
COE	Cops of Engineers
CO_2	Carbon Dioxide
CSDS	Chemical Specification Data Sheet
DIV 2000	Dutch Intervention Values (2000)
DO	Dissolved Oxygen
EC	Electrical Conductivity
ECRA	Environmental Clean-up Responsibility Act
EPA	Environmental Protection Agency
EPA ID	Environmental Protection Agency Identification Number
EQA	Environmental Quality Act and Regulation
ESA	Environmental Site Assessment
GC	Gas Chromatography
HRS	Hazard Ranking System
HSE	Health, Safety and Environmental
IPA	Iso-propryl alcohol
ISO	International Organisation for Standardisation
LOR	Level of Reporting
MIC	Methyl isocyanate
na	not analysed
nd	not detected
NESHAP	National Emission Standard for Hazardous Air Pollutants
NGO	Non-governmental Organisation
NPL	National Priorities List
PCB	Polychlorinated biphenyls
PDC	Penang Development Corporation

PID	Photo-ionisation Detector
PRG	Preliminary Remediation Goals
PRP	Potentially Responsible Party
PVC	Polyvinyl Chloride
QA/QC	Quality Assurance and Quality Control
RCRA	Resources Conservation and Recovery Act
RIVM	Dutch National Institute for Public Health and the Environment
RL	Reduced Level
RM	Ringgit Malaysia
RPD	Relative Percent Difference
SAMM	Skim Akreditsi Makmal Malaysia
SARA	Superfund Amendments and Reauthorisation Act
SVOC	Semi Volatile Organic Compounds
SWL	Standing Water Level
TCLP	Toxicity Characteristic Leaching Procedure
TNB	Tenaga National Berhad
VOC	Volatile Organic Compounds
UK	United Kingdom
US	United States
USCS	United States Classification System
USEPA	United States Environmental Protection Agency
WWTP	Waste Water Treatment Plant

CHAPTER ONE

Introduction

Soil is one of the familiar material that we take for granted. It is important for the survival of human race as it provides the basic ingredients and support for the growth of arable crops, grassland, trees, which provide food, fibre for clothes, and timber for buildings and fuel. We depend on the earth together with water, air and radiation from the sun to provide the essentials for life. The awareness to maintain soil fertility has been emphasized since early days as the farmers have considered soil as a resource on which their livelihood depended.

The development of higher yielding crops and the extensive usage of fertilizers, pesticides and irrigation have led to the over production of food, including animal products. The call by governments and Non-Governmental Organisations (NGOs) is to use less intensive practices. The public is concerned about the quality of food, drinking water and air. It has thus led to criticism of the use of fertilizers and pesticides. With the government's emphasis on industrialization, there are more manufacturing industries being built to earn revenue and reduce unemployment. The industries may practice some cost reduction programmes or activities to remain competitive. For example, they may discharge untreated process water to the soil or river at their locality or dispose hazardous waste in an unapproved manner, thus contaminating the site or water stream.

In summary, pollution occurs when some part of the environment is made harmful to living organisms. Leading to contamination being the influx of hazardous material to the soil, groundwater or air with the potential of hurting any living organism. For example, insecticide which is used to kill locusts would not be considered a pollutant unless it also kills other organisms that are considered to be beneficial to the environment. Most chemicals are harmful in concentrated form and may be beneficial in low doses. Zinc is an essential element for both plants and animals but is harmful to both in high concentrations.

The Environmental Site Assessment (ESA) process focuses on the evaluation of a site. The process outlines the characteristic of the condition of the soil, groundwater and material on site. Countries have started developing their own standards to conduct ESAs. The International Organisation for Standardization (ISO) started developing an ISO standard and was completed in 2004. The standard was formalised as Soil Quality Characterization of Soil related to groundwater protection (ISO 14015, 2004) and is getting international acceptance.

Motivation

This Environment Site Assessment study has been carried out on an industrial site. It was a one and a half storey detached factory. The purchaser wanted to protect himself from liabilities with regards to any environmental contamination on the site by the seller. By doing so, the purchaser conducts a due diligence Environmental Site Assessment (ESA) on the site with permission from the seller.

This practice is not common in some countries as the Department of Environment has not made this process a prerequisite in the sale of a property. Neither is there any legal regulatory requirement for such practices. Although it is a common process in the United States and in some countries in Europe.

The buyer being an established multinational, was required by the corporate headquarters to conduct the due diligence ESA. This activity was to protect the corporation from inheriting a liability in the process of purchasing or leasing a property. Upon completion of the study, the buyer will be in a position to quantitatively make a decision whether to go ahead with the purchase of the property. They would also know about the liabilities they will be inheriting and how to manage the outcome in due course. The intent is also to establish a base line for the soil and groundwater of the site. The study can also be shared with the Department of Environment. This action will establish a transparent relationship between each other as well.

In most developing countries there are no legal regulatory requirements requiring the buyer or seller to conduct an environmental site assessment. As such, this is not a common practice, as there are no prerequisite requirements established. So this exercise will benefit all affected parties. The process will also provide visibility to the condition of the investigated site. The local environmental agency can use the study as a reference to establish a protocol in the future.

Objectives

The objective of this book is to assist readers and researchers in conducting an ESA of sites or projects undertaken by their team. Students will be able to follow through the process to complete projects, reports or thesis. Also, for the general audience to understand of how the process of due diligence is applied. The intent of the book is to accomplish the following objectives:

- An assessment to establish the validity of the proposed protocol/manual.

- An assessment of contamination level of the site to determine the liability associated with it. It will help the buyer to decide whether to buy or not buy the property.

- To review operational practices conducted at the site and where possible, current and historic practices at surrounding facilities to assess the potential of contamination to the site under study.

- To conduct sampling of paint used at the site. The paint used had any lead content in it.

- To assess materials used on the site with the potential of having fibres of Asbestos Containing Material (ACM).

- To assess potential on site health risks as a result of identified soil and groundwater contamination at the site. This will allow the evaluation of potential environmental concerns and liabilities to the site.

- To conduct an assessment of the offsite activities that could potentially lead to contamination of the site.

- To recommend a strategy for remediation should risk based screening levels exceed the internationally recommended level.

- A baseline for soil and groundwater can be determined on the basis of the data obtained from the investigation carried out. The data can be used for future improvements or litigation matters with the Department of Environment.

Scope

The scope of the study is limited to the site under investigation and will include any offsite contamination from surrounding sites. They are limited to the following:

- Assessment and identification of potential sources of contamination by performing a walk-through audit of the site.

- Identification of any potential historic onsite contamination.

- Identification of any potential historic off site contamination sources from adjacent locations of the site.

- Identification of locations points for boring of wells for the environmental site assessment of the site.

- Drilling and installation on site wells to conduct the study.

- Screening the soil samples for volatile organic compounds (VOCs) using a photo ionization detector (PID).

- Collection of samples for soil.

- Collection of soil Quality Assurance sample.

- Sampling for volatile organic compounds (VOCs) in the soil.

- Sampling for semi volatile organic compounds (SVOCs) in the soil.

- Sampling for metals (arsenic, cadmium, chromium, lead, selenium, silver, barium, and mercury) in the soil.

- Development of groundwater monitoring wells - gauge and purge monitoring wells prior to collecting ground water samples.

- Collection of samples for groundwater Quality Assurance.

- Sampling for analysis of volatile organic compounds (VOCs) in groundwater.

- To collect sample for analysing semi volatile organic compounds (SVOCs) for groundwater.

- Sampling for analysis of metals (arsenic, cadmium, chromium, lead, selenium, silver, barium, and mercury) in groundwater.

- Assessment and tabulation of soil and ground water quality levels against Dutch Intervention Value (DIV 2000).

- To provide comments on the data, draw conclusions and recommendations.

Choice of Methodology

The scope of the study is limited to the site being investigated. Under normal circumstances the in main intent of the buyer is not to inherit environmental liability from the transaction. As such, this research study will need to fulfil the basic requirement of the stakeholder. Also, the study will need to follow the methodological steps of a case study. The intent of the study has to take an overall review of the site and provide enough information for the stakeholders to make realistic decisions.

Galliers (1992) defined case study as an attempt of describing relationships, which exist in reality within a group. The site being investigated is located in an area which is homogenous and can only be characterized by conducting a study with a systematic approach. The data of the site will be very specific to the location and comparing them to established available standards will allow the researcher to draw conclusive results. Garman and Clayton (1997) gave a similar definition of a case study. They described case study as "an in-depth investigation of a discrete entity". This means the investigation can be of a single setting, subject, collection or event. They also mentioned that it would be possible to derive knowledge and or a decision from the intensive investigation of the subject matter. The buyer's intent is to obtain sufficient information, so that they can make a calculated realistic decision in purchasing the mentioned property.

Reliability is always an issue in any research or study. Essentially, reliability concerns the extent to which a research will yield the same or similar outcome if the research is repeated. The reliability of the test data will be verified with spiked and duplicate sampling from the site. This process will enhance the reliability of the data with regards to collection and test equipment. The contents and scope can be verified with existing prescribed methodologies. The enhanced prescribed methods in the United States are scoped in the American Society for Testing and Materials (ASTM) E1527 and E1528 standards published in 2000.

Benefits

Conducting a study or research on any subject matter will enhance knowledge of the researcher. In this case, the researcher is applying all the theoretical skills, personal experience on environment, health and safety practices over the years, using some of the guidelines and established protocols used by existing western developed countries.

The benefits of the research or study can be summarized as follows;

- An assessment of the site will be obtained.
- The buyer and seller will each know their liability prior to the sale materializing.
- The research study will indicate the extent of environmental degradation of the site.
- The research study will indicate the risk implication with regards to health and safety to the occupants of site and neighbours.
- It will help in establishing the appropriateness and usage of the site of a responsible citizen.
- This practice of determining the environmental and safety of the site and correcting the site of the pollutants will create an emphasis of a green environment.
- The outcome of the research can be shared with the Department of Environment. This will generate knowledge sharing between the researcher, buyer, seller and the local environmental agency.
- The study comes with a protocol/manual for the assessment of sites, which can be used to assess sites for investigation and or abandoned sites.

CHAPTER TWO

LITERATURE REVIEW

Introduction

Many countries regulate the identification and remediation of contaminated sites. Whereas developing countries do not have any specific laws governing the requirements with regards to identification and remediation of contaminated sites. However, in recent years developing countries are starting to take interest and are empowering the local environmental agencies to take action against polluters.

Among the countries, United States takes a lead in defining requirements, establishing laws and implementing them with regards to polluters. Liability for the clean-up of these sites usually lies on the polluters and if they cannot be found or held liable, the current owner or occupier of the site is held responsible for the clean-up. Thus, to protect innocent owners, the government has introduced and regulated the transfer of contaminated and potentially contaminated sites. Therefore, when acquiring properties, a due diligence investigation is conducted to avoid any future environmental liability. Rules governing this process are updated and enhanced periodically for the benefit of purchasers.

Due Diligence Audits

Transactions involving real property or through company acquisitions can convey the liabilities associated with contamination present in the soil or ground water. Bankers who finance the property place their secured capital at risk if collateralised property is contaminated.

Certain states such as in New Jersey, in United States require a due diligence audit as per Environmental Clean-up Responsibility Act (ECRA, 1984) to be conducted prior to the transfer of an industrial property. The liability (with regards to cost) associated with remediation of a contaminated property is very huge and has surprised many land owners, tenants and lending institutions.

Some examples quoted in ECRA (1984) are discussed below:

A purchaser bought a piece of land and finds out later, the piece of land was used as a landfill. The purchaser had to pay US$6 million to clean-up the groundwater in the area. An electronic manufacturing company moves from a site that they had leased for 6 years. The site was acquired by another company 8 years later. Two years later, groundwater contamination was discovered. The site owner successfully sues (10 years later) the acquiring company to pay for the clean-up, claiming that the manufacturer was the 'only tenant likely to have used a sufficient quantity of the chemical known to have caused the contamination'.

A bank lends money to a company. When the business started to have financial difficulties, the bank stepped in and ran part of the company. The bank ultimately foreclosed the company. The property was found to be contaminated. The bank was found to be liable for a major part of the remediation. The cost of remediation was 25 times the amount of the original loan.

Thus, the value of real property can be greatly diminished if the property is found to be contaminated and needs remediation. The need to know the potential liabilities is crucial to both the buyer and seller of the property. The costs of removal of asbestos insulation, or remediation of soil and groundwater contamination resulting from historic uses can greatly exceed the present time value of the property. Lending institution may lose the property's value as collateral and possibly become liable for the clean-up cost upon fore closure. For the above reasons it has become prudent that all buyers and commercial lending institutions require an evaluation of the environmental risks before completing the mortgage transactions. Today in the US, due diligence audits have become a routine part of commercial real estate transaction. Some American and European companies operating in Malaysia have made it mandatory for its subsidiaries to conduct due diligence audits prior to any purchase or leasing of real estate.

American Society for Testing and Materials Standard Practice E1527

The industry standard for conducting due diligence audits of real property is provided by American Society for Testing and Materials (ASTM). The goal of the standard is to identify 'recognisable environmental conditions' that could lead to or result in Comprehensive Environmental Response, Compensation and Liability Act (CERCLA, 1980) liability at a site. What this means, is 'the presence or likely presence of any hazardous substances or petroleum products on a property under conditions that indicate an existing release, a past release, or a material threat of a release of any hazardous substances or petroleum products into structures on the property or into ground, groundwater or surface water of the property'.

The standard was further enhanced and is now known as the ASTM E1527-2000 Standard. The scope expansion includes:

- Interviewing of current owner or occupants,
- Interviewing with an owner of a neighbouring property,
- Examination of documents,
- Documentation of significant gaps or uncertainties and
- Specifying documentation requirements where the subject property cannot be visually inspected.

Parties interested in a Due Diligence Audit

In most cases, parties involved in a due diligence audit are a buyer and a seller. Parties can be an individual, several people and/or a corporation either way. The due diligence audit once performed is beneficial to both the buyer and seller to come to terms during the negotiation process. Each of them will know their individual liabilities from the results of the audit and will assist in the negotiation of setting the price. The audit process will assure the financial institution the value of the real estate and it also indicates that the property is free from any major environmental liability. Especially in an industrial or commercial property, the tenant and landlord should evaluate liabilities, prior to occupying the property and also on leaving it. This activity will avoid any future claims. There are no regulatory requirements in Malaysia with regards to conducting a due diligence audit. As such, the Department of Environment will not require a due diligence audit during the transaction of a property. Whereas regulatory agencies in the United

States have the authority and part of its objective is to protect the environment may bar the transfer of a property to its new owner until an audit is completed and/or the site remediated.

Elements of Audit

The focus of a due diligence audit is on historic practices and current operations that may result in the release of hazardous substances. Based on the ECRA reports, the largest liabilities derive from historic rather than current practices. The uncertainty of defining the liabilities is the difficulty in detecting the environmental problems. Some examples are unrecorded waste disposal operations, use of contaminated fill material and leaks in underground storage tanks.

For an industrial site, a through audit is required and some of its common elements are listed below:

- Investigation in soil and groundwater contamination.
- Presence of hazardous waste and substances.
- Industrial waste water discharges from treatment plant or septic tanks.
- Review and compliance records of approved permits for water, air and material.
- Assessment of the presence of underground tanks, waste storage areas, pits, sumps and floor drains.
- Usages of PCBs on site.
- Usage of lead based paints on the site.
- Usage of material containing asbestos.

Superfund

Comprehensive Environment Response, Compensation and Liability Act (CERCLA) Superfund was established in 1980 with a fund of US$1.6 billion. Its primary objective was to implement a massive environmental clean-up programme in the US over a five-year period. Generators were required to report to the Environmental Protection Agency (EPA) any facility at which hazardous waste are generated, or have been generated, treated, stored, or disposed. The aim was to identify and clean up hazardous waste sites first and then to litigate to recover the cost. The combination of unrealistic expectation and a deregulation attitude had the impact of slowing down an already difficult programme. At the end of the five-year period only six sites could be cleaned up.

CERCLA was amended in 1986 by the Superfund Amendments and Reauthorisation Act (SARA) 1996. SARA created a US$8.5 billion fund for cleaning up abandoned waste disposal sites and an additional US$500 million for cleaning up leaking underground petroleum tanks. This programme allowed the EPA flexibility to perform removal actions, implement more stringent state standards and establish a preference for permanent remedies that reduce the volume, toxicity and mobility of toxic substances.

During this period the Bhopal tragedy (1984) occurred in India. So, the 'Community Right-To-Know' provisions were included in the SARA. This required industries to plan for emergencies and inform the public of hazardous materials and substances being used.

Once a site has been identified, EPA uses the Hazard Ranking System (HRS) to estimate the degree of risk each site possesses and its impact to human health and the environment. Factors such as proximity to population, nature of the contamination, potential release and migration pathways are evaluated and then combined to get a HRS score. When the RS score exceeds the specified threshold value, the site is proposed for placement on the National Priorities List (NPL). After the site gets listed on the NPL, EPA sends notices to the identified Potentially Responsible Party (PRP), giving them opportunity to perform the required studies and perform the clean-up. The PRPs are present, and they are the past owners, operators of the facility at time of operations, generators and transporters. If no agreement is reached, EPA will perform the clean-up work through the US Army Corps of Engineers (COE), outside consultants and contractors working for the COE. Upon the clean-up work being completed, the EPA sues the PRPs to recover the cost spent from the fund.

There are nine criteria used for the selection of remedies of the site are as follows:

- Overall protection of human health and the environment.
- Compliance to clean-up standards.
- Long-term effectiveness and permanence.
- Reduction of toxicity, mobility, or volume of treatment.
- Short-term effectiveness.
- Implement-ability.
- Cost (capital, operational and maintenance).
- State acceptance.
- Community acceptance.

Clean-up Standards

How clean is clean? This is the most difficult question when it comes to evaluating contaminations in soil or ground water. What standard should be applied, what level is the contaminant and how can it be removed? It is not possible to remove contamination down to the last molecule. Thus, in 1986 the Act was amended and it required remedial actions to meet 'Applicable or Relevant and Appropriate Requirements', also known as the ARAR. There are three types of ARARs, namely:

- Chemical specific - levels are indicated for each chemical.
- Activity specific - landfills will have the appropriate standards.
- Location specific - disposal is prohibited at flood prone areas.

When an ARAR do not exist for important chemicals at the site or where EPA determines an environment regulation is not appropriate to determine clean-up levels, then such levels may be set through the use of quantitative risk assessment.

Quantitative Risk Assessment

Risk assessment is a way of evaluating potential hazards arising from contaminated lands. The hazard is a substance that has the potential to cause harm to a receptor, such as causing adverse health effects, groundwater unfit for use or damage to underground structures. Stubbs and Tang (2004) in their paper mentioned that contaminated sites have the potential of releasing chemical vapours. These vapours can accumulate in any enclosed space or environment and when the concentrations are high enough they can pose a short term threat of fire and or explosion. When the concentrations are low it will or may cause long term health risk to the occupants.

The UK requirements took a lead role and it took them three years to come up with a risk based approach to contaminated land management and in the recent years this approach has been promoted at a European level known as Contaminated Land Rehabilitation Network for Environment Technologist (CLARINET) in 2001. The basic principal to risk based approach to contaminated land management are not new, as there is mention of risk assessment by the Public Health of Scotland Act 1897.

The aim of risk assessment is to ensure that the land is fit for either its current use or its intended use. As such, the land needs to be assessed so that all living habitants are not put to risk.

As mentioned in the Statutory Guidance of the Scottish Executive (2000), the risk assessment relies on investigating the presence of the following;

- Presence of a pollutant linkage,

- Significance of a pollutant linkage and

- Establishing a relationship between a contaminant and receptor by a pathway.

All three elements of the linkage must be present for a risk to be present. It needs to be proven that the pollutant linkage can result in causing harm to the receptor. Or the pollutant linkage can cause significant pollution of controlled waters or soil. There are possibilities of more than one pollutant being present at a contaminated site. Having established the presence or likely presence of a risk, a risk assessment need to be conducted to determine the significance of the pollutant linkage. The risk assessment could be either qualitative or quantitative.

Qualitative Risk Assessment - involves the development of a conceptual model and identifying the potential pollutant linkages. There are basically five steps in this model and they are as follows:

- Desk Study - use the available documents and draw up a plan indicating likely sources of contamination.

- Walkover - comprises a plan, cross-section, text and network diagram.

- Site investigation - add details to the plan developed earlier.

- Monitoring - report on changes and indicating changes with respect to time.

- Remediation - add information to reflect changes.

- Quantitative Risk Assessment - are assessed based on data quality compared to decreasing conservation. These methods may be sequenced as follows:

- Compare site data with generic assessment criteria,

- Compare against site specific assessment criteria and

- Investigate and compare extended data collection of site with data collected from earlier investigations.

Most of the guidelines do not indicate a structured methodology of risk assessment. It is mentioned to maintain the principal of a stepwise approach. Ideally the assessment should be more site specific, with decreasing uncertainty and over conservatism.

In determining whether the site is contaminated the investigator and the local authorities need to know and establish the presence of the following:

- Contaminant - harmful substance or material.

- Pathway - the identified contaminant is causing harm to the receptor or there is a possibility that the contaminant will cause harm to a receptor. That the contaminant is causing pollution or likely to cause pollution to controlled waters.

- Receptor - can be human beings, controlled waters (fishing rights), ecological systems, property (crops, produce grown domestically for consumption, livestock, wild animals and buildings).

There are commercially available risk assessment tools which can be used for site investigation purposes. Some example of tools such as Contaminated Land Exposure Assessment (CLEA) is used when the receptor are people. When the receptor is groundwater, the ConSim V2 (2006) risk assessment can be used. When the receptor is both people and groundwater then the tools Risk Based Corrective Action (RBCA). The tools mentioned are common and there are many other tools available commercially for risk assessment.

Bhopal Gas Tragedy

The Bhopal gas tragedy happened on the 3rd of December, 1984 on Sunday at 1:00 AM. Shrivastava (1987) reported that approximately 45 tons of methyl isocyanate (CH_3NCO) leaked from a storage tank at the Union Carbide India Ltd., pesticide plant in Bhopal, India. Diamond (1985) called this 'the worst chemical accident in human history', nearly 3000 people were dead and about 200,000 injured.

Methyl isocyanate (MIC) is a main chemical used in the manufacture of pesticides. This chemical is highly unstable and has to be kept at low temperatures. It is extremely toxic, capable of causing severe Broncho-spasm and asthmatic breathing when inhaled. It is also an irritant and can be absorbed through the skin. Sax and Lewis (1989) indicated that exposure to high concentration of this chemical can cause blindness, damage to lungs, emphysema and ultimately death.

The cause of the accident has been ascribed by the media to be the flaws of the design, operating deficiencies, maintenance failures and inadequate training. Union Carbide's internal investigations suggest that the accident was the direct result of employee sabotage. Kalelkar (1988) presented and reported that the detailed investigation conducted by A.D. Little Inc, suggested that the likely cause of the accident was due to the direct entry of water into the MIC tank. The mixture of water and MIC produced high temperatures and pressures causing the MIC gas to be generated at an extreme high rate.

Due to the outbreak of this incident the US EPA was criticised because it had not listed MIC as a dangerous pollutant. Also, the magnitude of the problem led the US Congress to include the provision 'Right-To-Know' in SARA.

Cradle to Grave Concept

The USEPA has implemented a cradle to grave concept for its hazardous waste management. Hazardous waste is tracked from its generation point (referred to as the cradle) to its ultimate disposal site (referred to as the grave). The system requires all generators to attach a manifestation form to all shipments of hazardous waste. This tracking method allows the monitoring and control of all hazardous waste within the state or across state boundaries. The tracking of hazardous waste is managed by the RCRA regulatory act. Each generator is assigned a unique identification number for the waste that is being generated. The generator is not allowed to treat, store, dispose and transport any such waste without the EPA ID. Also, transporters of waste are regulated and required to be licensed. Appropriate packaging is defined so that the contaminant does not spill during transportation. The generator is required to generate a contingency plan and provide emergency procedures in case of an unforeseen situation. Personnel handling the waste, be it the manufacturer, processor or transporter are provided appropriate training. In line with the above, the Malaysian Environmental Quality Act (EQA, 1974) also requires generators to comply with similar guide lines.

Qualified Site Assessors

In the United States the CERCLA establishes the qualification for site assessors. The standard defines an environmental professional as one who must supervise the project of due diligence, prepare and sign the written report of the results.

The environmental professional in the US is to be suitably qualified. The list below is the minimum requirements:

- Holds a current Professional Engineer's or Geologist's license and have the equivalent of three years of full time relevant experience, or

- Licensed or Certified to perform environmental inquiries as defined by the standard and have the equivalent of three years of full time relevant experience, or

- Has a Bachelors or higher degree from an accredited institution of higher education in science or engineering and the equivalent of five years of full time relevant experience, or

- Having the equivalent of ten years of full time relevant experience.

Similar qualification requirement is the basic need to accomplish a due diligence investigation report. The above qualification may be used as a guideline when developing countries start to develop their own standards.

The Contaminated Land Management Act 1997 of the State of New South Wales in Australia requires site auditors to be accredited under the Act to conduct an independent review that relates to investigation, or remediation carried out in respect of the actual or possible contamination of land. The intent of the Act is as listed below:

- To define the nature and extent of the contamination of the land.

- To determine the nature and extent of the investigation or remediation.

- To specify the suitability of the land use.

- To specify the investigation or remediation necessary prior to the land is suitable for use.

- To determine/assess the suitability and appropriateness of a remediation plan, a long term management plan or a remediation proposal.

Other countries that have introduced regulatory systems to address soil and groundwater contamination include the United Kingdom, Netherlands, Germany, Belgium, France, Spain and Canada.

As such, companies in the United States wishing to acquire sites should be aware of what they should do to obtain and keep this protection. They must use an appropriately qualified environmental professional to perform the due diligence investigation.

Asbestos Containing Material (ACM)

Asbestos used in building materials was widespread because of its strength, chemical resistance, fire resistance and applications as acoustical and thermal insulation. In the early 1970s, asbestos use began to decline following the publication of various health studies. A particular study worthy of mention is the study done by Dr. Irving Selikoff (1991) at the Mount Sinai School of Medicine in New York. This study revealed that inhaling asbestos fibres has been shown to cause asbestosis (a fibrotic lung disease), lung cancer and mesothelioma (a form of cancer).

Estimation of asbestos related health risk is complicated because asbestos related diseases sometimes do not occur for 20 to 30 years after the material is inhaled. Also, other factors increase the health risks. For example, the risk of lung cancer is greatly increased when asbestos exposure is coupled with cigarette smoking. The health risk and the extent of exposure in an industrial environment are considerably higher than that of a commercial building. Studies have shown that the size of the asbestos fibre is important. It is now believed that the fibres less than 5 microns (a micron is one millionth of a meter) in length are not biologically as active as the longer asbestos fibres. Based partly on this assumption, Occupational Safety and Health Administration (OSHA) regulates only those fibres longer than 5 microns.

To minimise the risk level and ensure a safe environment it is recommended to include an ACM survey when conducting a due diligence survey. The objectives of the survey are as follows:

- Identification of suspect asbestos containing material,

- Bulk sampling of suspected asbestos containing material,

- Perform analysis of samples collected in accordance with the procedures specified, such as the United States Environmental Protection Agency (USEPA) at an accredited laboratory, and

- Review of site layout drawings to mark the location of ACM.

ACM is a good insulator and they are commonly found in thermal insulation and surfacing materials. Thermal insulation will include pipe wrappings, fittings and elbow joints, duct insulation, heat shields and insulation on chillers, heaters, boilers and air conditioning units. Surfacing materials will include fireproofing material, acoustic plaster, decorative plaster, spray applied insulation, floor tiles, ceiling tiles, transite board, vibration damping material in ducts and wall board. Areas where it is difficult to check for asbestos is like gaskets, covered flooring material, concrete block and press wood products.

Lead Based Paint

Lead is a toxic chemical when digested or inhaled. The effect it has on humans can be very damaging. Waldron (1966) reported the impact of lead on human and is shown in Table 2.1 in summary. The study reports that the epidemiologic studies on human exposure to the toxic substance have convincing evidence of having adverse human health effects. The data is of cases limited to occupational exposures and to single catastrophic exposure incidents involving the general population. The United States Department of Health and Human Service's agency for Toxic Substances and Disease Registry (1988) has listed that the level of lead in blood of 15 to 25 micrograms per decilitre can have serious implication relating to central nervous system functions.

Table 2.1: Short and Long Term Effects of Lead Concentration to Human

Effect on human	Lead Concentration in Blood (µg/dL)				
	1 to 9	10 to 18	19 to 30	30 to 100	≥ 100
Short Term ≤ 14 days		Possible effect on infants from mother's exposure during pregnancy		Brain and kidney damage	Death
Long Term ≥ 15 days	Inhibition or precognitive development in children below 6 years	Possible effect on infant from mother's exposure during pregnancy	Increased in blood pressure in middle aged men	Brain and kidney damage	Death

Short term exposure of 10 to 18 µg/dL to pregnant women can have effect to the foetus. So long term exposure can have severe adverse effect not only to the foetus but also to normal man-kind.

Toxicity Characteristic Leaching Procedure (TCLP)

This is a procedure for extracting sample and performing a chemical analysis. This procedure was adopted by the US EPA in November 1986. This method of analysis is a regulatory requirement. The main purpose of this test is to evaluate the effectiveness of stabilization of the contaminant in the soil. The US EPA uses the TCLP (1990) for classification of materials as hazardous or non-hazardous.

The stabilized material is crushed to small particle sizes (less than 9.5 mm) and is mixed with an extraction fluid at a pH of 2.88 which is made up from water and acetic acid. This is to mimic the condition in a municipal landfill. The ratio of the extraction fluid to the material is 20:1 and the mixture is agitated for 18 hours at 30 rpm and 22 degrees centigrade. After the extract is filtered through a 0.7 μm filter and defined as the TCLP extract. It is now tested for a variety of hazardous waste such as VOC, SVOC, and metals. The results of the analysis are than compared to regulatory requirements as stipulated in the 40 CFR 262.24 Federal Register.

The TCLP testing can simulate a situation that can occur in a field such as a landfill. The main concern is the migration of the contaminants into the groundwater stream and continues the track into the environment. That is why it is important for hazardous waste to be solidified and stabilised prior to landfill. There after continue to monitor for leachates on a periodic basis as a control measure. Malaysia in the last ten years has set up landfills to accommodate for waste to be treated and stabilised.

Construction of Monitoring Wells

In most countries, groundwater is an important resource, as it is the source of drinking water. In some countries it is the only source. It is crucial to plan for the use of groundwater for the present, future and to understand the quantity and quality of water that is available.

By measuring groundwater table elevation, it is possible to determine the direction of groundwater movement. Once the groundwater flow regime is known and it will be possible to trace the associated contaminant distribution and transportation.

As such, the construction of the monitoring well plays an important role in this assessment. Thus the two important principal uses of the well are:

- To determine the water flow hydraulic head and
- Allow sampling of water quality for measurement purposes.

Monitoring wells are similar to water wells. They are used to measure movement and monitor quality of groundwater. Most monitoring wells are finished above ground surface to make them easy to work with and easy to find.

Surveying is a method of measuring precise distances in horizontal and vertical directions to a datum. Mean sea level is the mean elevation of the water in the earth's seas. It is used throughout the world as the point of reference for all elevation measurements. To determine the movement of the groundwater, measurements are taken from monitoring wells at different locations. Well variations, such as well depth, ground surface elevation, groundwater table depth, give hydro geologists the necessary data to calculate groundwater movement in the area. It is common practice (Chapelle, 1993) to conduct drilling operation using hand augering, auger drilling or air drilling for the investigation of soil or groundwater in subsurface soil. This service is commercially available and the equipment is not costly. In cases for evaluation of marine environment (Smith et al. 2000) different techniques are applied. The costs for such evaluation are much higher compared to terrestrial grounds. The cost is high due to investment of equipment and skill of crew doing the investigation.

Pressure head in the subsurface is accomplished via a piezometer which is explained in the Appendix 3.

Soil and Groundwater Sampling

For investigation of soil and groundwater sampling of wells there are two approaches, either a professional judgement or targeted sampling. In most cases, a combination of the two may be used. As indicated in standard CLR4 (DoE 1994), sites are investigated to determine average concentration and standard deviation at a 95% UCL (Upper Confidence Level) for mean. In normal statistical evaluation the recommended sample size is 30. But when it comes to soil or groundwater analysis the sample of 30 may not be feasible and will depend on the preliminary results of the first few wells and outcome of interviews conducted of the site. Thus, a professional judgment is determined based on value of site, suspected contaminants and locality of site.

Based on the interviews conducted it was determined the site is very unlikely to be contaminated. The interviewed personnel did not disclose or highlight any knowledge or observed any foreseeable risk of soil contamination due to spills or during handling of chemicals. Also, the chemicals used were appropriately stored on the production lines and in the warehouse. Chemical on the production area were stored on spill skids and the warehouse or storage area there was a containment pit. The containment pit volume was large enough to contain two large (44 gal per drum) drums if spillage was to occur. The Malaysian legal requirement for the containment pit is 110% of the largest drum or container.

The other reason for limiting the number of wells to 13 for the investigation is cost. Each additional well will increase the cost of investigation, development of the well and analysis cost of soil and groundwater.

Based on professional judgment, preliminary data of the first few wells and result of statistical analysis data of the first few wells the investigation was limited to 13 wells. Sample well locations for the investigation are selected based on available information of site, site audit and review. The intent is to investigate an area or location suspected to contain contaminant.

CHAPTER THREE

METHODOLOGY OF APPROACH

Phase One Study – Introduction

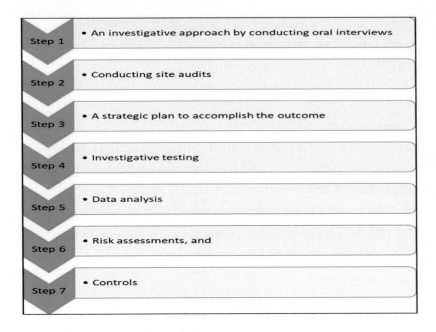

Step 1	• An investigative approach by conducting oral interviews
Step 2	• Conducting site audits
Step 3	• A strategic plan to accomplish the outcome
Step 4	• Investigative testing
Step 5	• Data analysis
Step 6	• Risk assessments, and
Step 7	• Controls

This case study research reported in this thesis constitutes of planned activities that were taken through a systematic approach and verification on the reliability of data. The flow was cyclic from following steps 1 to 7 and in continual momentum. At times it took a few cycles to accomplish an outcome.

This methodology has been supported by several researchers. Feagin and et al (1991) have used this approach and received much attention from its users. Several users such as Yin (1993, 1994), Stake (1995) and Tellis (1997) have used this methodology in several of their research. With their wide experience in this methodology, they have developed robust procedures in this area. Thus, if the researcher follows the prescribed methodology, then the researcher is following methods that are well developed and tested as any in the scientific field. This research is based on experimental work and adhering to procedures in the protocol or manual used in this study. Also, as Tellis (1997) mentioned, the case study approach is made to bring out the details from the viewpoint of the participant by using multiple data.

The Environment Site Assessment (ESA) will be able to identify obvious environmental contamination, safety and health hazards associated with the property. In ESA the investigated property is assessed for visible signs of possible contamination. Property records of the site are reviewed with the local authorities and discussion are held with regulatory agencies. Past and present employees are interviewed to obtain information and practices on site. The management staffs and owners of neighbouring sites are interviewed. This information will enable the researcher to draw a conclusion to provide an appropriate recommendation to the purchaser of the property. Each topic in Table 3.1 below describes the needs and specific requirements for the research to be completed in the report. Further most of the methods are written in the imperative sentences because they are intended to be used as reference manual by others in their research as well as assessment studies. It also fulfils the need of the topic of this thesis for the protocol development. The audit protocol or manual shown in Table 3.1 will be used in this study as it was developed from the literature review of the available and acceptable standards available. It is more comprehensive than any individual standard.

Table 3.1: Elements of Audit Protocol/Manual

Audit Protocol/Manual	
Flow	**Activities**
Phase 1	Purpose and Limitation Site Description Records Review Agency Contact and Investigation Previous Studies if Available Property Description Hazardous Materials Drums and Storage Containers Solid Waste Water Supply and Waste Water Discharges Potential PCB Containing Equipment Underground Storage Distressed Vegetation
	Drilling (3 wells)
Phase 2	Asbestos Lead Based Paint Investigation Surrounding Property Reconnaissance Hydrogeology Hydrology Drilling Soil Sampling Groundwater Sampling Project Cost
Phase 3	Remediation

Purpose and Limitation

The intent and scope for the investigation is described under this topic. An identification and traceability track of the property is assigned to explain the type of land status (freehold, leased, commercial, residential or industrial). This is because the value of the property will vary depending on the status and purpose. A property classified as a residential site cannot be used for commercial or industrial purposes. The information can be obtained by conducting a search at the land office or reviewing the title of the property.

Site Description

Clearly identify the site location on a map. This will provide a clear location plan of the investigated property. It will aid the reviewer in locating the site without much difficulty. The reviewer will also be able to visualise the site on the map and layout. Where possible a location map to be provided from a macro to micro view.

- To provide site description indicate the neighbouring properties on each side or direction of the investigated site. Also indicate the characteristics, what they do and practice. This being an industrial site it is crucial to report on the final product and what are the raw materials used. If the neighbours are willing to share, then also indicate the waste streams that exist on their site.

- To describe the closest residential site, indicate rivers or water catchments areas in the vicinity and also indicate the closest airport and port of the investigated site.

Records Review

This is a documentation trail to establish ownership and materials used on the site. Research is conducted on the materials used to determine its characteristics and identification of those that are hazardous. Most of the details can be obtained from the Material Specification and Data Sheet (MSDS). In Malaysia this documentation is referred as the Chemical Specification and Data Sheet (CSDS). It is required for the following:

- To review the owner's record with regards to the title of the property and describe the operational status and the intent of selling the property.

- To review the business products registered and details of the raw materials purchased that were used and quantities consumed over the period of occupancy.

- To provide a flow chart describing the manufacturing flow, where possible, which will assist the reviewer to visualise the manufacturing process by reading the report.

Agency Contact and Investigation

Establish contact with the local land office to determine the status of the property. It will indicate the number of owners and the status of the land. The local agency will be able to provide documents for the approval obtained for business operational practice. The records may be in the form of maps and/or written records. It is needed for the following purpose:

- To verify land title documentation with the land office.

- To conduct enquiries with the land office on any future development plans in the vicinity or on the investigated site. This is to provide transparency from the legality prospective.

Previous Studies if Available

Contact the previous owners to obtain information whether any previous study has been completed on the site. This will be a good starting point on determining contamination that exist or have been remediated. It will be helpful for the following:

- To consult the seller whether they had conducted any previous site investigation and or environmental site studies.

- To verify the report if it is available with regards to potential contaminant on site or from any potential contaminant from off site.

Property Description

Provide an overall macro status of the property being investigated. These documentations can be in the form of aerial photographs, photographs of the site, building plans and/or sketch plan. They will be helpful in the following:

- To describe the physical size of the investigated property, neighbouring properties and vacant land.

- To describe any constructed buildings existing on site and condition of the buildings.

- To indicate the current practices of the factory.

Hazardous Materials

Material contamination is a common source on sites. As such review hazardous material purchased, used or generated on site. A review of hazardous material disposal methodology is a good starting point and will provide the necessary information for the following:

- To describe any hazardous material that is on the site, based on physical presence during the interview and audit process.

- To review for distressed vegetation at material storage areas and any oil stain marks on the site.

- To review the disposal documentation to disposal site and records submitted to the Department of Environment, if waste is identified.

Drums and Storage Containers

During the audit process, observe for drums and storage containers. They are an indication that some form of storage has occurred. It will help:

- To review for any containers lying around on the site.

- To describe the storage methodology of containers.

- To audit for the presence of secondary containment at storage site.

- To review the condition of the drums.

Solid Waste

This waste is commonly disposed at the municipality council waste dump. They are collected by the council workers on a regular basis depending on the generation of waste. Items can be paper, cardboards, non-hazardous waste from the production and discarded office material. This will determine whether hazardous materials are also disposed illegally or out of ignorance. If hazardous materials are disposed through this channel, there can be future implication if contamination is found at the municipality council dump site. It is done for the following reasons:

- To identify location where solid waste is stored.

- To review the site and the content of the waste.

- To describe the waste content and also verify whether the waste is correctly characterized.
- To describe the house keeping around the storage area.

Water Supply and Waste Water Discharges

Most industries use water in their processes especially for plating and cleaning. Chemicals are also used in the cleaning process, thus leading to discharge water being contaminated. It will be advantageous for the following:

- To indicate how the site gets its water supply?
- To indicate the number of wells on site and its uses.
- Describe how waste water is discharged from the process, wash area and rain water.
- To review discharge water characterization and records. Where no records are available to collect samples and send to accredited laboratory for analysis.
- To indicate any discoloration of the drain and any odours from the drain or around the site.

Potential Poly Chlorinated Biphenyl Containing Equipment

Poly chlorinated biphenyl (PCB) was commonly used in capacitors or transformers as a dielectric media since the 1950s. Mackenzie and Cornwell (1991) indicated in their text that PCB ceased to be in production in 1977 in US. It is compound with low water solubility and has a very long half-life in soil. The half-life is estimated to be about ten years in the subsurface and at least one to three years on the soil surface. PCBs can also accumulate in the sediments of bodies of water. Due to the low solubility of water and in the absence of photochemical degradation they tend to persist in the sediments or accumulate in the bottom dwelling organism. Follow the steps given below:

- Identify all potential PCB containing equipment on the site.
- Verify the dielectric media of the equipment by looking at the test data and CSDS (Chemical Specification Data Sheet). Observe for any spills, odour or discoloration of the site where the equipment is located.

Underground Storage

Underground storage tanks provide a unique environmental problem. Their very presence is unknown to people living in the vicinity until a problem occurs. When a problem occurs it is too late for preventive measures as the contamination would have happened. The common causes of contamination with underground tanks are corrosion or leakages of the tanks.

To prevent future leaks, owners of underground storage tank need to provide either a leak detection system or an inventory control with regular testing of tanks. The tanks need to be tested for corrosion on a regular basis as well. It will serve the following purposes:

- To indicate any storage tank on site.
- To specify what chemicals or material that is stored.

- To specify if the tank is above ground or underground.

- To describe and verify the controls undertaken to prevent spillage or leakage.

- To survey the site surroundings.

Distressed Vegetation

During the site audit make observation for distressed vegetation. This is a visual indication of contamination having occurred. The site that has an area with limited plant growth need to be identified for further investigation. Water and soil analysis need to be conducted at the localised area for contamination for the following:

- To observe for distressed vegetation during site audit.

- To identify and indicate the size of vegetative area and covered area.

- To describe the boundary area with respect to the neighbour's site.

- To observe for any distressed vegetation around the boundary area. Affected areas will have limited growth of grass or plants.

Condition of Surface Soils and Soil Disturbances

During the site audit review the condition of the soil surface and soil disturbances. The soil surface will indicate whether the soil has been conditioned with soil from another site especially if the site is a low lying area. If the soil has been disturbed such as excavation work has been done, ensure water and soil samples are taken at the localised area. Following are the aims of this audit:

- To describe the surface of the soil on site.

- To indicate any soil disturbances on site.

- To look out for any discoloration, oil stains or spill on the soil surfaces during the walk through audit.

Air Emissions and Odours

Air emissions may occur in a variety of forms at process level. The typical way of characterisation of air emission sources are instantaneous. Emissions may come from a chimney, stack, vent or any opening. Emissions can be in the form of particulate matter, organic, metal and can also be in gaseous state. Combustion processes can release gases as a by-product or due to incomplete combustion and particulate matter in the form of fly ash or not combusted carbon.

Most of emissions may contain an odour. In the walk through audit process the need to be watchful for odours around points of discharge. Also conduct tests for soil and water around the point of discharge. This survey is needed for the following:

- To identify air emission points on site and indicate on layout.

- To review the permits and conditions for operating the emission point.

- To review the emission results with respect to country requirements and conditions stipulated in the conditions of operations in the permit.

- To review the site of the emission point and verify the site for any contamination, discoloration or odour.

Phase Two Study

Asbestos

Asbestos is a common material used in buildings and in insulation of piping. Likely areas of asbestos used are in pipe insulation, wall panel, floor tiles, ceiling tiles/panes, insulation material for oven doors and sealing tapes. Asbestos fibres are linked to cause cancer of the lungs if inhaled in large amounts or for a long period of time. Its audit is conducted for the purposes mentioned below:

- To identify asbestos suspect material.

- To collect bulk samples in accordance with criteria set forth in the USEPA Asbestos Hazard Emergency Response Act 1986, (AHERA) regulation.

- To analyse samples collected for asbestos content at an accredited laboratory.

For the purpose of the current study, the sample points have been identified and are discussed as below.

Since the building was empty, there were limited number of sources for the investigation and detection for asbestos fibres. Nevertheless, the following sampling points were identified. The samples points were identified based on potentially risk materials used at the site, such as gypsum panel, ceiling panel, vinyl floor tile and wall panel. Sample A1 was from the office wall panel on the ground floor. Sample A2 was from the ceiling on the 1st Floor of the office area. Sample A3 is from the 1st Floor ceiling in the canteen area. Sample A4 is the floor tile from the production area. Sample A5 is from the wall panel in the production area. Sample A6 is from the ceiling in the MDN coil production area.

The samples taken are tabulated in Table 3.2 below. The last column of Table 3.2 also indicates the rational for the selection of the sample. Figure 3.1 indicates the location points of the asbestos samples taken. The samples were taken based on the criteria mentioned above.

Table 3.2: Location Points for Asbestos

Location ID	Location of Sampling	Rational – Assessment
A1	Ground Floor office wall panel	Potential asbestos exposure from gypsum wall pane to office workers
A2	1st Floor office ceiling	Potential asbestos exposure to maintenance of facilities personnel and office environment
A3	1st Floor canteen ceiling	Potential asbestos exposure to maintenance or facilities personnel and employees using the canteen
A4	Production floor tiles	Potential asbestos exposure to employees working in production area
A5	Production wall pane located at the west end of the facility	Potential asbestos exposure to employees working in production area
A6	Production ceiling	Potential asbestos exposure to maintenance or facilities personnel and employees working in the production area

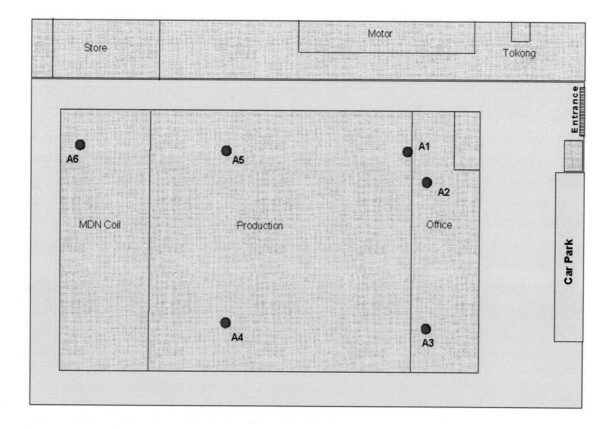

Figure 3.1: Location Points for Asbestos Testing

Lead Based Paint Investigation

Exposure to lead contamination may cause severe illnesses even to the extent of death. In the US, lead paint is strictly regulated in paints for household purposes and in children's toys. Finkel (1983) and Waldron (1966) indicated that lead poisoning causes critical illnesses, such as anaemia, kidney tumours and inhaled lead leads to paralysis. Lead exposure is monitored and governed by regulations. As such it is crucial that sources of such contamination are thoroughly investigated. The four step approach undertaken is as follows:

- To identify potential areas of lead contamination (to follow US regulated guidelines). To collect samples of paint from various locations of different colours. Identify areas where paint work is fraying, then pick one to two samples of each colour.

- Sample collection should be conducted by a qualified industrial hygienist, personal protective equipment to be worn during the collection process, the need to comply with appropriate protocols for sample handling.

- Samples collected to be sent for analysis to an accredited laboratory.

- Review results from the laboratory and provide corrective action as appropriate.

For the purpose of this study, after the walk through audit the sampling points were selected. Eight sampling points were identified for sample collection. They were numbered from Pb-1 to Pb-8, as can be seen from the Table 3.3 below. It describes the identification of the sampling point, specifies the location and also describes the rational for selecting the sample. The points were selected based on paint colours. The point of selection was based on sampling convenience.

Sample Pb-1 is grey and is from the door of the transformer room. The paint was slightly lighter and maybe due to direct exposure to sunlight. Sample Pb-2 is brown in colour and is from the handrail leading from the office area to the 1st floor. Sample Pb-3 is orange in colour and is from the handrail leading from the canteen to the office floor. Sample Pb-4 is grey in colour and is from the front wall of the front side of the building. Sample Pb-5 is red in colour and is from the water pipe to the hand wash area of the canteen. Sample Pb-6 is dark brown in colour and is from the main door frame at the front entrance. Sample Pb-7 is brown in colour and is taken from the ladder at the rear of the building. The ladder leads the way to the roof of the building. The brown is also slightly faded and can be due to exposure to direct sunlight. Sample Pb-8 is light grey in colour and is taken inside the building next to electrical panel.

Table 3.3: Location points for Lead in Paint Sampling

Location ID	Location of Sampling	Rational - Assessment
Pb – 1	Door of transformer room located at the west side of building	Potential lead exposure to TNB personnel and facility maintenance personnel
Pb – 2	Hand railing from office at ground floor to 1st floor	Potential lead exposure to office and housekeeping staff
Pb – 3	Hand railing from office at ground floor to canteen (on 1st floor)	Potential lead exposure to employees and housekeeping staff.
Pb – 4	Wall near front door at east side of building	Potential lead exposure to employees and housekeeping staff.
Pb – 5	Water pipe on 1st floor of canteen	Potential lead exposure to employees, maintenance and housekeeping staff
Pb – 6	Main door frame of front entrance at east side of building	Potential lead exposure to office and housekeeping staff
Pb – 7	Ladder located at the west end of the building	Potential lead exposure to maintenance staff
Pb – 8	Wall of production floor next to the electrical panel	Potential lead exposure to employees and housekeeping staff

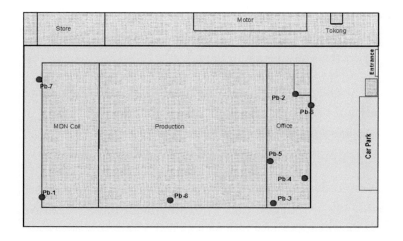

The layout of Figure 3.2 indicates the sampling points taken for lead based paint investigation.

Figure 3.2: Layout indicating Sampling points of Lead based paint Investigation.

Surrounding property Reconnaissance

Information collected about the neighbouring surrounding properties can provide information with respect to any external influence. There have been cases where the underground water stream has contaminated neighbouring sites. By knowing the chemicals utilised by the neighbouring sites it may simplify the investigation process. This can be achieved by conducting oral interviews of the neighbours. It will be helpful in the following:

- To describe the surrounding properties and area.

- To indicate what is manufactured at the location and any potential contamination from the adjacent sites that could influence the investigated site.

- To conduct interviews with the adjacent site occupiers, with regards to the operations, chemicals in use and waste streams.

Hydrogeology

Hydrology is the study that deals with the properties and movement of water. The continuous circulation of water in the atmosphere, in the subsurface and in surface water bodies and is termed the hydrological cycle. The cycle begins with precipitation of water that drops on the earth's surface in the form of rain. The water may spread in the form of surface water runoff or may seep into the ground and become groundwater. Throughout the cycle water may enter the subsurface environment through infiltration or return to atmosphere through evaporation. Water reaching the saturated zone in the subsurface will flow from areas of high hydraulic head pressure to areas of lower hydraulic head pressure. The subsurface layer that permits this water movement is termed as an aquifer. An illustration of the above is provided graphically in Appendix 4.

The groundwater will move dissolved contaminants caused due to spills and leakages of storage containers. The contaminants will disperse with the flow of groundwater. In some cases, if the situation is not corrected immediately the cost to remediation may be high and also cause risk to living organisms.

In the subsurface environment, contamination transfer is dependent on groundwater flow. Hydrology is the study of properties and flow of water and geology is the study of soil properties and characteristics. This exercise will provide information and data in the process of investigation and experiments conducted. To optimise cost it will be beneficial to conduct the investigation in steps.

First, conduct preliminary investigations by drillings wells on a macro basis to determine any major contaminants on site. If any previous study has been conducted and is available, that could help in determining location of wells and points of contaminations. Also, information from offsite investigation conducted from interviews, chemical inventory analysis and audits.

Based on results from borehole 1, 2 and 3 it was deduced that further investigations was required. It was necessary to drill 10 more wells surrounding the investigated site. This will provide more conclusive and reliable results in the investigation process. Samples were selected based on potential risk areas.

The rational and assessment in selecting the location of the wells is in the third column of Table 3.4 below.

Table 3.4: A brief description of each well location.

Well ID	Location of Well	Rational – Assessment
Bore hole BH1	Southern corner of facility	Potential impact from septic tank
Bore hole BH2	South west site boundary	Potential impact from offsite sources
Bore hole BH3	North west site boundary	Potential impact from offsite sources
Bore hole BH4	Southern corner of facility	Potential impact from offsite sources
Bore hole BH5	Southern eastern corner of facility	Potential impact from septic tank
Bore hole BH6	Southern corner of facility	Water quality at site boundary
Bore hole BH7	Southern corner of facility	Water quality at site boundary
Bore hole BH8	South west corner of facility	Potential impact from offsite sources
Bore hole BH9	Western boundary	Potential impact from offsite sources
Bore hole BH10	Northern boundary	Potential impact from offsite sources
Bore hole BH11	Northern boundary	Potential impact from offsite sources
Bore hole BH12	Coil manufacturing area – some thinner recycling was done at this location	Potential impact beneath factory – from historic onsite operation
Bore hole BH13	Production area – assembly of products	Potential impact beneath factory – from historic onsite operation

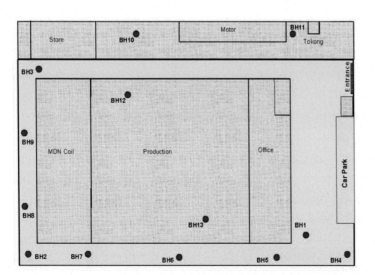

Figure 3.3 identifies the position of the well on the site. The circle indicates the well and it is labelled with a borehole number. This site layout will assist in the identification of the wells.

Figure 3.3: Site Layout and Monitoring Well Location Plan

26

Hydrology

This exercise will define the groundwater flow, water cycle and contamination transportation. The rain water will spread along the ground surface as surface water runoffs and seep into the ground and become groundwater. The water will flow underground to streams and lead to the river reaching the sea. The water underground will flow from the higher to lower hydraulic head, thus moving contamination underground through the soil to the sea. The water flow is derived on Davey's Law (Mackenzie and Cornwell, 1991). Contamination flow is based on the condition and type of soil. Sandy soil generally allows contamination to move faster compared to clay like soil. When the soil is clayey (clay like) contamination tends to be localised, meaning movement is restricted. It is required for the following:

- To describe the hydrology of the site.

- Indicate any stream, river or shore line surrounding the site. The water movement underground can at times influence the site positively or negatively. This only affects sites located near coastal areas because the tide can influence the site.

Drilling

If site conditions permit the use of hand powered auger then it will be preferred choice for the drilling. The intent is to drill up to 5 metres underground or 1.5 metres below the water table. Hand held augers consist of an auger bit, a solid or tubular rod and a 'T' handle. When the drill rod is threaded, extensions can be added or auger bits interchanged. The auger tip drills into the ground as the handle is rotated. The soil retained on the auger tip is brought to the surface and used as soil sample. Alternately, augers can be used to bore to the desired sampling depth, and a tube sampler used for collection. A wide variety of auger tips are available to suit a variety of soil types. The preferred form of hand auger to be used for the collection of soil samples should be that which can take a core sample.

This was possible because the results from the first three holes provided conclusive evidence that the water level is about one metre below ground level. This method has the advantage of collecting undisturbed soil samples and it is also suitable for investigating areas where safety related issues may restrict access of other techniques. It is also suitable to check for the presence of services under ground. This method is safe to use around piping and utility lines.

The groundwater monitoring wells were designed and constructed in accordance with internationally recognized protocols for environmental investigations. The wells were constructed using a 50 mm diameter PVC blank casing. The casing had a 0.3 mm aperture, horizontal machine slotted screen with flush treaded joints. The intent of having a well screen is to allow the entry of potentially free floating and dissolved organics. The annulus of each well was filled with a filter pack comprising clean sand to about 0.2 metres above the top of the screen. A layer of bentonite was then placed above the filter pack to prevent the surface run off from entering the well. After which each well was plugged with a well plug. This is to protect any external contaminants from entering the well.

For permanent wells a casing will enclose the opening of the well at just below the ground level. The casing is used to protect the well and also ease future evaluation of groundwater at the site.

Soil Sampling

Soil samples are taken during the drilling of the wells. Soil samples are collected as per the ASTM prescribed methodology.

- To collect soil samples where contamination is observed or the possibility of contamination is suspected.

- To collect the samples at the level of the water table, unless the water table is very deep such as greater than 20 meters.

In this study the soil samples were collected during the process of constructing the borehole. The soil samples collected were tested for VOCs (volatile organic compounds), SVOCs (semi volatile organic compounds) and metals. VOCs are compounds that have a high vapour pressure that allows them to evaporate easily. Examples of VOCs are benzene, toluene, styrene and vinyl chloride. Whereas, SVOCs includes a broad band of compounds with different physical and chemical properties. Examples of SVOCs are phenol, pyrene and chrysene (a polycyclic aromatic hydrocarbon). This will allow corrective action to be taken and conduct risk assessments for the site.

Groundwater Sampling

Upon drilling, installation and developing the wells, the groundwater was collected as per the ASTM prescribed methodology. This will allow the well to be stabilised and the groundwater sample will be representative of the water and condition of the aquifer being studied. Refer to Appendix 4 on monitoring well for some explanation on water table and aquifer. In order to ensure and avoid areas of floating free phase product, install the intake of the screen one metre below the groundwater level. If the ground of the site being tested is compact and free from floating free phase, then the intake screen can be located one metre below the ground level. This will prevent any cross contamination between samples.

- To install groundwater monitoring wells both within and around the site boundary. The boundary wells are used to check on the quality of the groundwater that enters and leaves the site. This will allow an assessment on contaminated groundwater entering or leaving the site to be made. Groundwater collected from the wells are tested for VOCs, SVOCs and metals.

- Maintain notes during the drilling, installation, development and sampling of well with regards to water levels, colour, type and log the data in the borehole log. Complete the data in the table of the log.

Project Costing

For such type of project there is involvement of cost involved in accomplishing the necessary investigations and in researching per the scope outlined. The cost of investigations will vary depending on the location and characterisation of the land under investigation. Nevertheless, it is beneficial to know the quality of the purchased plot as it will prevent future liabilities. The site had a transaction value of one million three hundred thousand ringgit. The cost for investigation was about ten percentage of the transaction value. The detail costing of the project is shown in Appendix 9.

The plot of land under investigation is zoned as an industrial land and it is not close to any drinking water streams. As such, the testing for risk assessment will be limited.

Quality Assurance and Quality Control (QA/QC) Procedures

In order to ensure that the sampling protocol is being executed appropriately and the situation which might lead to errors are recognised before they seriously impact the results of analysis it is necessary to undertake various QA/QC procedures. These procedures involve the use of various blanks, duplicates and matrix spikes. This will enable quantitative corrections for bias which arise due to handling, storage, transportation and laboratory procedures and identification of possible of cross contamination between samples and or equipment.

It is essential to conduct an analysis of the samples that were taken with regards to the reliability of the results from sampling and testing of the results.

Phase Three Study - Remediation

During the course of the study and investigation, if any contamination is detected a remediation process to remove or contain the contaminant is to be provided based on risk assessment on the site occupants.

CHAPTER FOUR

RESULTS AND DISCUSSIONS

Phase One Results – Introduction

It is expensive to conduct site assessments, as such it is appropriate that caution is applied in the planning process to minimise expenditure. The property is assessed for visible signs of possible contamination; records of local authorities are reviewed interviews with employees, interviews with management staff, discussion with regulatory agencies and owners of neighbouring sites. The information gathered during the research will be useful in planning for the investigation process.

Purpose and Limitation

This is a study of an Environmental Site Assessment (ESA) being conducted on one and a half storey detached factory complex erected on the leasehold land. The site is located on a 5900 square metres plot and is a gazetted industrial zone. This subject property was purchased from the local government agency under a leasehold condition for 60 years.

This ESA will be able to identify obvious environmental contamination, safety and health hazards associated with the property.

Site Description

The site is located in an industrial zone estate. On the site, a one and half storey industrial factory is located. The site is located among other factories in the zone. The size of the site is not as large as some other factories in the locality. The site is accessible through the main highway to most major cities.

There are several properties adjacent to the site being investigated, which are as follows:

- A glass factory located on the northwest of the site. This plant manufactures glass optical lens and prisms.

- A ceramic's factory is located on the northeast and it manufactures ceramic capacitors. North of the ceramic factory is an electronics factory.

- On the southeast of the site is a public road. Opposite the road is an electronic factory which manufactures a wide range of electronic components and one of them is high reliability capacitors.

- On the southwest of the site, the land is vacant. There is a large concrete drain that leads the drainage water into the sea.

- South of the drain the factory manufactures latex household gloves and nitrite synthetic rubber gloves.

There is a large residential area located about 2.3 Km east of the site. The airport is about 20 kilometres from the site. It takes about an hour's drive during non-peak hours. Peak period is defined as time during shift changes of the factories. The times of shift changes are around 7 AM, 3 PM and 11 PM. The traffic in the region is also heavy during the office hour changeovers at 8 AM and 5 PM. The port is located about 7 kilometres from the site and takes about 30 minutes' drive from the site.

Records Review

The owner stopped operations and decommissioned the site in 2004. The owner is currently residing in Taiwan. The product that was manufactured at the said site has been transferred to their China factory. The reason quoted for the transfer was due to the unavailability of local direct labour force, resulting in escalating labour cost. Most of the factories are depending on foreign direct labour, especially from Indonesia, Bangla Desh and Vietnam. Due to the closure of the factory it was not possible to verify all the site documentation. To overcome this issue, interviews with ex-employees and the neighbouring factories was conducted. Based on the limited data and information from the interviews an outline of historical operational practices was established for the site. The questions and responses (in italics) from the interview are shown in Appendix 11.

Agency Contact and Investigation

The land had belonged to the local government board and was leased to the seller. The land has been classified as an industrial site and the seller was licensed to manufacture on the site. The ownership of obtaining the appropriate approvals to operate the factory lies with the owner.

Previous Studies if Available

The seller informed that there were no prior environmental site assessment studies conducted for the site. This is a common practice followed by most of the factories that started operations in this area. There was little need as this area started developing into an industrial zone about fifty years ago. This locality used to be an agricultural area. This was verified with interviews conducted at the land office for the region. The property was purchased from the local government by the seller.

Property Description

The site area is approximately 5,900 metres square. The land is generally flat and is about 0.5 to 1 metres higher than the surrounding properties. This site was used as a factory for the production of ceramic capacitors and coil inductors. According to the Geological data obtained from the local authorities, the site is underlain by quaternary marine and continental deposits of clay, silt, sand and peat with minor gravel.

Hazardous Materials

There was no hazardous material observed on the site during the audit process. This could be due to the production has been relocated to their China plant. During the interview process it was determined from the ex-production manager that all related raw materials used in the manufacturing of the product, line tooling and the manufacturing equipment were also relocated to China. In the building, the floor was clean and no materials were found during the audit and inspection. No oil stains were observed on the floor. Most of the space was covered with concrete in the building and the external was covered with asphalt. There is a small garden area in front of the building with a pond.

Drums and Storage Containers

There were no drums or storage containers found on the site that contained any hazardous materials or chemicals. There were no stains observed on the floor where chemicals and raw materials were stored. The housekeeping at the site was good.

Solid Waste

There was no solid waste on site. The owners had cleaned up the site upon relocating the process to China.

Water Supply and Waste Water Discharge

Potable water supply for the site is obtained from the local municipality. There were no groundwater wells observed on the site. Also, there was no waste water treatment plant operated by the previous owner. This was concluded from the oral interview sessions with ex-staff of the facility and site review. The building is surrounded with storm water drains and the exit of the drain leads into the sewage septic tank. Rain water runoff is channelled to the storm water drain. There were no stains observed in the drains and did not exhibit any chemical odour.

Potential Poly-Chlorinated Biphenyl Containing Equipment

Electrical transformers and capacitors are a potential source of environmental concern due to the potential presence of polychlorinated biphenyls (PCBs). Some make or units of transformers and capacitors contain dielectric fluids which may contain PCBs. The national electrical power company substation is located on site. Interview with the ex-Production Manager on site indicated that the substation used mineral oil as the dielectric media and was PCB free. As such, there were no records that could be verified on site. The substation is maintained by electrical power company.

The site had its own electrical distribution on site. The electrical equipment used mineral oils as insulation fluid, and they are PCB free according to information from the Production Manager. The CSDS (MSDS) was verified and it did not contain any PCB. Based on observation, there was no potential PCB containing equipment noted. There were no oil stains observed around internal and external floor vicinity of the sub-station.

Explosion proof lighting was used in the sub-station and it was protected with carbon dioxide (CO_2) fire extinguishing media. Electrical switches were also explosion proof rating.

Underground Storage

There were no underground storage tanks noted during the site audit. This was further confirmed during the interview process with the ex-staff.

Distressed Vegetation

The building structure was occupying majority of the build-up area. All surrounding area from the building is covered with concrete and bitumen. There is a small plot of 2 x 10 metres of flower bed area complete with flower plants and an aquarium with fishes. No stressed vegetation was observed on site.

Condition of Surface Soils and Soil Disturbances

Most of the surface area of the site is covered with concrete and bitumen, as described above. The surface area surrounding the building was observed to be free from any form of digging. The internal surface of the building looked even and no patch work was observed.

Air Emissions and Odours

Since the manufacturing process has been relocated to China, a comprehensive survey of the manufacturing process was not necessary. All the machinery and production line tooling have been removed and the previous production area was empty. Nevertheless, a walkthrough of the site was conducted and there was no odour in and outside of the building. No odour was noticed around the boundary of the site. No air emission points were observed from the building located on the site.

Phase Two Results

Asbestos

Six points were evaluated for asbestos fibres. All the evaluated samples were free from asbestos fibres. It is part of the process to conduct the investigation on asbestos fibres in the area. As asbestos fibres inhaled over a period of time will lead to carcinogenic tumours in the lungs.

None of the samples were fraying and the building material was also in good condition. Under conditions when the material is a suspect and is fraying it becomes critical that the material is removed by a qualified industrial hygienist and disposed to an approved disposal site for incineration. Not doing so, can lead to exposure of the asbestos fibres to employees working in the vicinity. Also, appropriate personal protective equipment need to be worn when removing or working on suspect material that may contain asbestos fibres.

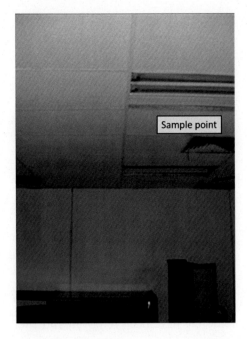

Plate 4.1 is location number A2 (Table 4.1) and is taken on the first floor of the building. The ceiling was analysed for asbestos.

Plate 4.1: Sampling point for Asbestos

Table 4.1 summarises the result of asbestos testing. Column three of the table indicates the level of asbestos content. No asbestos fibre was detected in all the samples that were analysed. Column four and five indicates the non-asbestos content and non-fibrous material of the samples analysed. Most of the samples analysed contained non -fibrous materials such as binder and filler material.

None of the samples analysed contain any carcinogenic fibres. No further investigation is needed on asbestos on the site.

Table 4.1: Results of Asbestos Testing

Location ID	Location of Sampling	Asbestos Content (% and Type)	Non-Asbestos Content (% and Type)	Non –Fibrous Materials (Type)
A1	Ground floor office wall panel	None detected	≤ 1% cellulose fibres	Binder and Filler
A2	1ST Floor office ceiling	None detected	≤ 10% cellulose fibres	Binder and Filler
A3	1st Floor canteen ceiling	None detected	≤ 35% cellulose fibres	Binder and Filler
A4	Production floor tiles	None detected	≤ 1% cellulose fibres	Binder and Filler
A5	Production wall pane located at the west end of the facility	None detected	≤ 1% cellulose fibres	Binder and Filler
A6	Production ceiling	None detected	≤ 20% cellulose fibres	Binder and Filler

Lead Based Paint

A total of eight samples were collected and analysed. Plate 4.2 is the hand rail from the office at ground floor leading to the canteen on the 1st floor. The sample was taken at the lower side of the railing. This sampling point is of location Pb-3.

Plate 4.2: Lead paint Sampling

Table 4.2 is the summary of the results of lead sampling. Results of lead concentration are listed in column four of the table.

Table 4.2: Laboratory Results for Lead Based Paint Investigation

Location ID	Location of Sampling	Colour of Paint	Lead Concentration (mg/kg)
Pb – 1	Door of transformer room located at the west side of building	Grey	200
Pb – 2	Hand railing from office at ground floor to 1st floor	Brown	180
Pb – 3	Hand railing from office at ground floor to canteen (on 1st floor)	Orange	870
Pb – 4	Wall near front door at east side of building	Grey	150
Pb – 5	Water pipe on 1st floor of canteen	Red	100
Pb – 6	Main door frame of front entrance at east side of building	Dark brown	120
Pb – 7	Ladder located at the west end of the building	Brown	80
Pb – 8	Wall of production floor next to the electrical panel	Light grey	50

MiniTab was used to analysis the significance of lead contamination in the paints collected from the site. Refer to Appendix 2 - Minitab Computation, for the method applied in the derivation of the analysis.

Refer to Appendix 1 - Statistics, for an explanation on hypothesis testing and evaluating the significance of P-value.

The following data and results were obtained from Table 4.3:

Null hypothesis, H_0: Lead in paint samples = 600 mg/kg

Alternative hypothesis, H_1: Lead in paint samples < 600 mg/kg

P-value obtained is 0.003 (0.3%)

The mean is 218.75 mg/kg

Standard deviation is 267.87 mg/kg

Table 4.3: Statistical Analysis of Lead Based paint

One-sample T-test:			Lead Contamination (mg/kg)				
Variable	N	Mean	St. Dev	SE Mean	95% Upper Bound	T	P
Lead (mg/kg)	8	218.75	267.82	94.71	398.18	-4.03	0.003

Graphical representation of the lead concentration is shown in Figure 4.4 and it shows that 7 of the data points are below the mean of 218.75 mg/kg of lead contamination. The null hypothesis value was taken as 600 mg/kg, this is the allowable limit for lead in paint in products for children based on the ASTM standard. There are no requirements or standards in Malaysia for lead in paints for domestic household use.

One of the data point shows a contamination point of 870 mg/kg. This is a hotspot for lead contamination. It is the orange paint that is used on the hand railing. It can be concluded from the data above that with a Confidence Interval (CI) of 95%, the probability of finding another point above 600 mg/kg is 0.3%, which is very low. The null hypothesis is rejected in favour of the alternative hypothesis, which is the value to be less than 600 ppm. Further investigation on this contamination is not necessary, as it would be very unlikely to find another hotspot of lead contamination.

The data point on the right side of the graph which is the out liner is the hot spot of the investigation. As can be seen all the other points are concentrated on the left side of the graph and are all below the sample average.

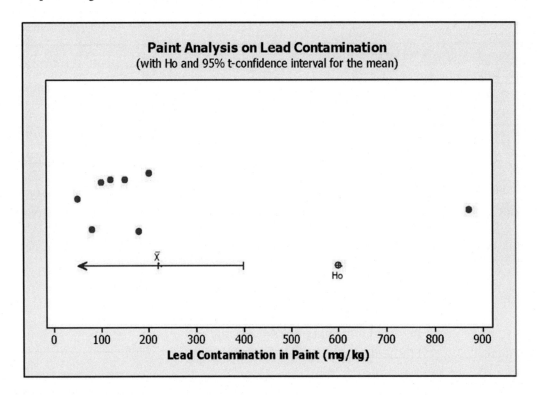

Figure 4.4: Graphical Analysis of Lead Based Paint

Surrounding Property Reconnaissance

The surrounding property reconnaissance was performed to assist in evaluating if adjacent land could have or would have contaminated the site. The site is surrounded with light industrial factories. There was no distressed vegetation observed around 100 metres of the site in all directions. The site direction is shown in Figure 4.5.

An oral facility interview was conducted with the human resources and facility department personnel of the neighbouring factories and the results are summarised as follows:

- The factory on the north side of the site manufactures ceramics.

- The factory further north is located two blocks north of the site. It provides sub-contracting services such as assembling products for a neighbouring electronics and the chemicals used on the facility are, IPA (iso-propryl alcohol) for cleaning purposes, solder wire, solder flux and thinner. All chemicals are stored on concrete flooring and there is no secondary containment. The plant does not generate any waste water as such it does not have a waste water treatment plant. There are no underground and above ground storage tanks at the facility.

- Another electronics factory is located three blocks up north of the site. It is currently operating as a trading facility. It used to manufacture electronic components such as termistor and varistor at the facility from July 1990 to April 2001 and during this period the facility had used thinner and alcohol.

- The factory on the west side of the site, manufactures optical blanks for camera lenses and LCD lenses. Chemicals used on at the facility during the manufacturing process are thinners, iso-propryl alcohol (IPA), emery powder, polymer and lubricants. The facility has a waste water treatment plant, caustic soda and acids are used in the treatment of process waste water. Sludge from the waste water treatment plant is stored on asphalt and bare ground about 10 to 15 meters from the site boundary. Waste glass is also stored about 5 meters from the site boundary. Both the sludge and the waste glass are stored on the west side of the site boundary closer to the south corner. Plate 4.3 shows the waste from the west side factory stored on the west side of the boundary more towards the south side of the site. The waste is close to the boundary and the potential of contamination from off site is possible. It was not possible to get the characteristic of the waste. Based from the interview with the employees it was determined that it is the solid part of the sludge form the waste water treatment plant.

Plate 4.3: Waste Stored the Factory on the West Boundary.

- The factory on the east side of the site manufactures V-clip and E-clip capacitors. The land is owned by a multi organization and has been on lease since year 2000. Chemicals used by the facility include alcohol and electrolytes. There is no waste water treatment plant. There is no underground tank and there is one 1000 litre tank above ground used as diesel storage tank. The tank has a secondary containment and is sheltered from rain.

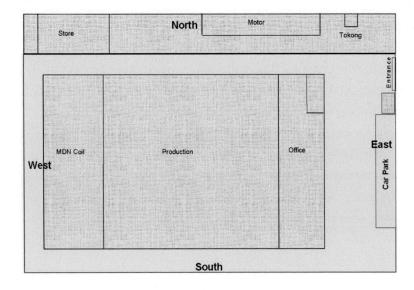

Figure 4.5: Layout plan of the Site being investigated.

Hydrogeology

Hydrology is the study that deals with the properties and movement of water and geology is the study of earth. Hydrogeology deals with the distribution and movement of groundwater in the soil and rocks of the earth's crust. The continuous circulation of water in the atmosphere, in the subsurface and in surface water bodies and is termed the hydrological cycle. The cycle begins with precipitation of water that drops on the earth's surface in the form of rain. The water may spread in the form of surface water runoff or may seep into the ground and become groundwater. Throughout the cycle water may enter the subsurface environment through infiltration or return to atmosphere through evaporation. Water reaching the saturated zone in the subsurface will flow from areas of high hydraulic head pressure to areas of lower hydraulic head pressure. As Deutsch (1997) mentioned that the flow of groundwater in subsurface areas are normally slow moving. The subsurface layer that permits this water movement is termed as an aquifer. An illustration of the above is provided graphically in Appendix 4. To determine the flow of groundwater we have software programmes that can provide the hydraulic gradient and one such programme is the Surfer. Some information is provided in Appendix 5 about Surfer.

According to Plummer and et al (1994), the transport of contaminants in the subsurface environment is dependent on the flow of groundwater. The groundwater will move dissolved contaminants caused due to spills and leakages of storage containers. The contaminants will disperse with the flow of groundwater. In some cases, if the situation is not corrected immediately the cost to remediation may be high and also cause risk to living organisms. Three soil bores were drilled using a cable tool rig. Plate 5 is showing the boring of bore hole number 1 (BH1) in progress. The maximum drilling depth was 5 meters below ground level. The water table was observed to be between 0.652 to 1.509 meters below ground level. The data will be discussed in later part of the text.

The site and the surrounding areas are supplied with piped potable water. Therefore, the use of groundwater as a potable water supply is considered unlikely. No groundwater wells were observed during the site review.

Hydrology

The nearest surface water body is the coastline which is approximately 850 meters south of the site. The tributary of a river is located approximately two kilometres east of the site. It joins another main river and flows in the westerly direction before joining the coastline.

Drilling of Wells

There was no layout drawing available on site for underground piping, drainage and electrical cables. As such screening of the well site position was conducted. The process uses a hand held sensor detector that screens for underground cables or piping. Information on underground cable detectors is in Appendix 3. The reliability of the equipment should be noted, some equipment is only sensitive to detect up to one foot deep only. So, as the construction of the well continues in depth the sensor detector should be progressively used for detection purposes.

All drill sites were cleared for underground electrical cables and services using a hand held cable detector. Each soil bore was drilled up to 1.5 meters below ground level depth using a hand auger. The subsequent required depth was achieved using the cable tool rig. An underground power cable was hit at approximately 1.3 meters below ground level for the first bore hole. Sparks and smoke was noted coming out of the hole. All work was halted and the power outage was reported to TNB. They requested that a police report need to be made prior to them repairing the damage. This could be due to the neighbouring factories may claim damages to their manufacturing processes. It is important to note that insurance should be taken when conducting such evaluations. The cost of repair to the electric cable and expenses charged by TNB was RM 5,800 (five thousand eight hundred Ringgit).

The borehole location was adjusted by a few feet from the underground electric cable. The reason the contractor was unable to detect the underground electric cable with the cable detector may be due its sensitiveness. Caution was used in constructing subsequent wells.

Plate 4.4 shows the site where the cable was damaged. The damaged cable was under the dark surface of the area. The area was dug 4 x 10 feet to a depth of 5 feet. The TNB technicians repaired the cable and covered the cavity with gravel. The top surface was covered with six inches of asphalt.

Plate 4.4: Completed patch-up with Asphalt after Electrical Repairs

The area was cordoned off for safety purposes. Time taken in drilling, preparing the well and collecting soil Plate 4.6 shows upon completion of bore hole BH1. The top of the well is covered with a cap to prevent any debris or contaminated material from entering. A metal plate was placed on the well for its protection. At a later date a well cover will be constructed around the well opening to make it a permanent feature.

Plate 4.5 - shows the crew working during the drilling of borehole BH1.

Plate 4.6: Completed Well for Borehole BH1.

Plate 4.7 shows the drilling of bore hole BH4. Since the surface area covered with concrete the crew are using an electric drill to cut through the concrete.

Plate 4.7: Concrete Coring of Borehole BH4 in Progress.

Plate 4.8 show the completed well of bore hole BH4. As can be seen in the photo the concrete is about seven inches thick. The well will be covered with a permanent casing at a later date.

Plate 4.8: Completed Well of Borehole BH4.

Plate 4.9 shows the crew drilling bore hole BH13 which is located inside the building. Since the cable jig cannot be installed inside the building the crew had to manually construct the well. The crew are using a hand auger in constructing the well.

Plate 4.9: Hand Augering of Borehole BH13.

Plate 4.10 shows the crew are in progress of drilling bore hole BH7. Since the well is located outside they are able to use the tripod cable jig in the construction. The 55-gallon drum is used to collect the soil that has been dug from the ground.

Plate 4.10: Drilling in Progress for Borehole BH7.

Soil Sampling

All soil borings were logged according to the United Soil Classification System (USCS). The profile of the boring was logged on a field sheet, soil properties, information and observation were recorded. Refer to Figure 4.6 for the sample of a bore log. The logs for all the bore holes are in Appendix 7 of the report.

Soil samples were placed in a glass jar with Teflon lined lid and labelled. The jars were then placed in a cooler box with ice. The samples were subsequently sent for analysis to the analytical laboratory which is an SAMM (Skim Akreditsi Makmal Malaysia) accredited laboratory. The samples were than tested as per the Table 4.4 below.

Table 4.4: Protocols Identified for the Analysis of Soil.

Soil Analysis	Analytical Methods	Sample of Soil	Preserved at Temperature (°C)
VOCs	USEPA 5030B, 8260B	250 mL	4
sVOC	USEPA 3570, 7270C	250 mL	4
Metals	USEPA 3050B, 6010B, 7471A	250 mL	4

Groundwater Sampling

The ground water monitoring wells were designed and constructed in accordance with internationally recognized protocols for environmental investigations. The wells were constructed using a 50 mm diameter PVC blank casing. The casing had a 0.3 mm aperture, horizontal machine slotted screen with flush treaded joints. The intent of having a well screen is to allow the entry of potentially free floating and dissolved organics. The annulus of each well was filled with a filter pack comprising clean sand to about 0.2 metres above the top of the screen. A layer of bentonite was then placed above the filter pack to prevent the surface run off from entering the well. After which each well was plugged with a well plug. This is to protect any external contaminants from entering the well.

Upon installation of the wells, they were purged three times to remove sediments and other impurities. They were then left for one day before sampling of ground water was taken. This is to allow the ground water to reach equilibrium level with the surrounding ground water system. All the wells were purged prior to sampling to ensure that any stagnant water was removed and only representative ground water samples were collected. Wells are normally considered to be adequately developed when a minimum of three well volumes of ground water were removed, field water quality indicators are stabilised or the well become dry.

Prior to the commencement of sampling activities, all wells were gauged for water table elevation. A hand held electric oil-water interface probe capable of measuring to plus minus 1 mm was used to gauge the wells. The reference point for the measurement was the top of the casing.

A designated bailer was used to purge each well and to collect samples. This is to minimize the risk of cross contamination between sampling locations. For sampling on dissolved metals analysis, the sample water was filtered in the field using a 0.45 micro meter filter unit.

All the samples were labelled, placed in ice and sent to the laboratory for analysis. The samples were tested as per Table 4.5 below. Table 4.5 also indicates the specified analytical methods and preservation for ground water testing. The reason for selecting the US protocol was that the buyer of the site is an American multinational company and the selected laboratory had the necessary experience in conducting the protocols specified. Based on the size of the job the laboratory also provided a technician on site to supervise the collection of samples.

Table 4.5: Protocols Identified for the Analysis of Groundwater

Groundwater Analysis	Analytical Methods	Sample Container	Preserved at Temperature (°C)
VOCs	USEPA 5030B, 82603	2 x 40 mL glass vials with Teflon lined lid	4°C, pH ≤ 2 HCl
sVOC	USEPA 3510, 8270C	1000 mL amber glass bottle with Teflon lined cap	4°C
Dissolved Metals	USEPA 6010B, 7470A	125 mL plastic bottle	4°C, pH ≤ 2 HNO3 Field filtered

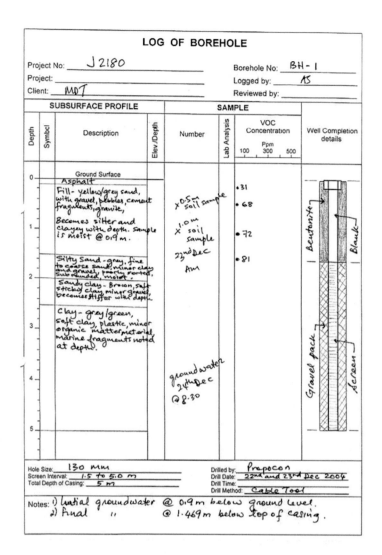

LOG OF BOREHOLE

Project No: J 2180   Borehole No: BH-1
Project: Logged by: AS
Client: MDT   Reviewed by:

SUBSURFACE PROFILE				SAMPLE			
Depth	Symbol	Description	Elev./Depth	Number	Lab Analysis	VOC Concentration Ppm 100 300 500	Well Completion details

Ground Surface
Asphalt
Fill- yellow/grey sand, with gravel, pebbles, cement fragments, granite,
Becomes silter and clayey with depth. Sample is moist @ 0.9 m.

Silty Sand - grey, fine to coarse sand, minor clay and gravel, poorly sorted, sub rounded, moist.
Sandy clay - Brown, soft sticky clay minor gravel, becomes stiffer with depth.
Clay - grey/green, soft clay, plastic, minor organic matter, pictorial marine fragments noted at depth.

x 0.5 m soil sample   • 31
x 1.0 m soil sample   • 68
22nd Dec   • 72
Ann   • 81

groundwater 24th Dec @ 8.30

Bentonite   Blank
Gravel pack   Screen

Hole Size: 130 mm
Screen Interval: 1.5 to 5.0 m
Total Depth of Casing: 5 m
Drilled by: Prepocon
Drill Date: 22nd and 23rd Dec 2004
Drill Time:
Drill Method: Cable Tool

Notes: 1) Initial groundwater @ 0.9m below ground level.
2) Final   "   @ 1.469m below top of casing.

Figure 4.6 shows a sample record of bore hole log BH1. As the bore hole is prepared the data is kept on record for future references and analysis. As can be seen on the log a description of the soil and various other data is kept with dimension and depth mentioned.

Figure 4.6: Sample of Data Record of Borehole Log BH1.

Project Costing

There are four types of cost involved when conducting site investigation projects. The cost headings are categorized as follows:

- Well installation,
- Testing of Components,
- Disposal cost related to investigation and
- Time spent on investigation, analysis and report writing process.

Well Installation

This is a skilled process and the availability of this skill is limited in Malaysia. In this investigation, the job was contracted out to a licensed and experienced company to construct wells on the site for the investigation. Since the company is located in about 180 Km away from the site, the first cost associated will be mobilization. This cost involved is to get the equipment needed for the investigation moved to the site. So, this cost will vary with proximity. In the case of this study the cost for mobilization was RM1725 (one thousand seven hundred and twenty five).

Prior to the installation of the well, the contractor will need to conduct an investigation to determine any underground cables or piping. Major of the cost is the premium for insurance purposes. The liability associated with causing damage to underground electric cable can be high and the potential of neighbouring sites may claim losses to their production schedule due to disruption of electricity supply. Depending on the job size the cost may vary from RM2000 to RM5000 (two to five thousand). This cost may be minimized if the site has a layout showing all the underground services.

Cost to install one well is RM1600 (one thousand six hundred) and this includes all the labour involved, material cost (pipes, sand, bitumen, etc), depreciation cost of the equipment and profit margin. An additional RM400 (four hundred) is charged to convert the temporary well to a permanent well. Temporary wells are used for short term purposes for the investigation and are removed after the process is complete. Whereas permanent wells can be used for longer period of time and the main reason to maintain permanent wells is to continue monitoring for longer period of time.

Testing Cost

This cost is related to testing of the components at the laboratory. The laboratory selected was a licensed and accredited laboratory. The criteria in selecting the laboratory is due to their:

- Accreditation and ability to conduct the testing as per the defined protocol,
- Turnaround time in providing results,
- Flexibility to collect samples from the site and
- Competitive pricing.

The normal lead time to get the laboratory results are about ten days, but in urgent cases the results may be provided in three days at a premium cost of 50% (fifty percent) above the published rate. The laboratory also provides a door to door service assisted by qualified laboratory assistants. This provided added reliability in the storage and transportation of the samples.

Cost to test for heavy metals is RM300 (three hundred) per sample, VOC (volatile organic compounds) is RM600 (six hundred) per sample and SVOC (semi VOC) is also RM600 (six hundred) per sample. For the testing of lead in paint and asbestos it is RM120 (one hundred and twenty) per sample.

Waste Disposal Cost

This is the cost related to the preparation of the wells. When the wells are drilled the soil is stored in large drums (44 gallons). These drums are then kept in storage until the laboratory results are reviewed. If there is no contamination than the soil may be disposed as domestic waste or reused at another location. If the soil is contaminated, then the soil is disposed to Kualiti Alam (KA) as scheduled waste for landfill (depending on characterization of contaminants in soil). Cost associated are containers, rental of handling equipment and disposal cost to KA. Most of the water collected from the wells during purging was transferred to a nearby site and flushed through a waste water treatment plant. Thus, there was no disposal cost associated for the disposal of waste water.

Time Spent on Investigation, Analysis and Report Writing Process.

Cost associated to this section is more of labour. The time spent in collecting data, performing interviews and conducting audits. Upon collecting all the information time was spent on analysing the data to draw conclusions. The data was used to perform statistical analysis, qualitative and quantitative risk analysis. This cost is not included in the costing. A detailed project costing is included in Appendix 9.

Quality Assurance and Quality Control Procedures

To ensure the reliability of the sampling and adherence to protocols several quality assurance and control procedures were instituted. The goal of this procedure is to ensure that the sampling protocol is followed faithfully and that the situation leading to errors are recognized before they seriously impact the data. The use of field blanks, standards and spiked samples can account for changes in samples which occur after sample collection. They provide independent checks on handling and storage as well as on the performance of the analytical laboratory.

Field blanks and standards enable quantitative correction for bias which arises due to handling, storage, transport and laboratory procedures. The spiked samples and blank controls provide the means to achieve combined sampling and analytical accuracy or recoveries for the actual conditions to which samples have been expressed.

Decontamination Procedure

All equipment used in the drilling and sampling process were decontaminated between sampling locations to minimize the risk of cross contaminations. Decontaminations procedures comprises of the use of phosphate free detergent wash and water rinse of sampling parts as appropriate.

The drilling equipment are also decontaminated between each bore hole. This is to prevent any gathering of contamination between sampling points of each bore hole. Material (such as screens, PVC casing, etc) used for the construction of wells are also decontaminated prior to installation.

During groundwater sampling the measuring equipment were also decontaminated between each bore hole to maintain water quality that is being measured and prevent any cross contamination.

Duplicate QA Samples Field Samples

Duplicate QA samples were collected to assess laboratory performance, identify any deficiencies and validate laboratory data. A field duplicate is an independent sample which is collected as close as possible to the primary sample. They are used to document the precision of the sampling and analytical process. Duplicates for this study comprised of one soil and ground water duplicate.

Samples QA - Laboratory

The laboratory is required to work to strictly according to QA protocols. Routine QA includes laboratory control samples, surrogates, method blanks, sample duplicates and matrix spikes. The results of samples are supposed to be compared to spike samples and the duplicate samples. This is to ensure that the equipment and the process used are within control and the results can be concluded as reliable.

Data Management and Reporting

The samples have to be maintained so that traceability is available. As such strict adherence is required in preparing and monitoring the samples. The QA samples are managed as follows:

- Sample labelling: Each sample is assigned a code describing the sample location, date of sampling and depth.
- Sample Transfer: Sample transfer protocol includes the COC records detailing the persons relinquishing and receiving the samples, sample descriptions, preservation details, turnaround time and analytical requirements.

Assessment

QA Assessment Results

Soil Analysis

Two QA samples were taken for soil analysis during the study, one duplicate soil sample at bore hole BH3 and another at BH5. Table 4.6 are the results from the sampling that were undertaken to verify QA data for soil analysis.

Table 4.6: QA Data for Sampling of Soil - Reliability of Sampling

Soil Analysis				Metals								
		VOC	sVOC	Ar	Cd	Cr	Pb	Se	Ag	Ba	Cr+6	Hg
Sample ID	Analysis						(mg/kg)					
LOR		-	-	5	1	1	1	10	1	10	-	-
BH3	Primary/ Primary Lab	nd	nd	nd	nd	0.60	42	nd	nd	9	nd	nd
Duplicate BH3	Duplicate/ Primary Lab	nd	nd	nd	nd	0.28	31	nd	nd	9	nd	nd
RPD between primary and duplicate sample (%)		-	-	-	-	73	30	-	-	0	-	-
BH5	Primary/ Primary Lab	nd	nd	nd	nd	2	94	nd	nd	40	nd	nd
Duplicate BH5	Duplicate/ Primary Lab	nd	nd	nd	nd	2	64	nd	nd	28	nd	nd
RPD between primary and duplicate sample (%)		-	-	-	-	0	38	-	-	35	-	-

Key
LOR – level of reporting
nd – not detected. Concentration below LOR
na – not analysed
RPD – Relative Percentage Difference

Groundwater Analysis

Two QA samples were taken for groundwater analysis during the study, one duplicate soil sample at BH1 and another at BH7. Table 4.7 are the results from the sampling that were undertaken to verify QA data for groundwater analysis.

Table 4.7: QA Data for Sampling on Groundwater - Reliability of Sampling.

Groundwater Analysis					Metals								
		VOC	sVOC	Ar	Cd	Cr	Pb	Se	Ag	Ba	Cr+6	Hg	
Sample ID	Analysis						(µg/L)						
LOR		-	-	50	1	1	10	100	10	50	50	0.2	
BH1	Primary/ Primary Lab	nd	nd	nd	nd	14	nd	100	nd	nd	nd	nd	
Duplicate BH1	Duplicate/ Primary Lab	nd	nd	nd	nd	32	nd	146	nd	59	na	nd	
RPD between primary and duplicate sample (%)		-	-	-	-	78	-	37	-	-	-	-	
BH7	Primary/ Primary Lab	nd	nd	nd	nd	38	nd	nd	nd	nd	na	nd	
Duplicate BH7	Duplicate/ Primary Lab	nd	nd	nd	nd	35	nd	nd	nd	nd	na	nd	
RPD between primary and duplicate sample (%)		-	-	-	-	8	-	-	-	-	-	-	

Key
LOR – level of reporting
nd – not detected. Concentration below LOR
na – not analysed
RPD – Relative Percentage Difference

QA Assessment

Data validation was carried out by calculation of relative percent difference (% RPD) between duplicate samples. RPD is the quantification of the accuracy of a measurement. RPD is calculated as follows; 'The difference between the two sample results divided by their mean and expressed as a percentage'.

Acceptable QA criteria for RPD measurement is as follows:

- RPD is <30%,
- When RPD is >30%, the primary and duplicate data are below the LOR,
- When RPD is >30%, the primary and duplicate data are 10 times greater than LOR.
- When one of the data is not detected (nd) and
- For blanks the criteria is not detected or below LOR.

Two duplicate soil samples were collected for borehole sample BH3 and BH5 and analysed for VOCs, SVOCs and metals. The Relative Percent Difference were within acceptable range for all the components that were analysed with the exception of Chromium in BH3, where the % RPD was 73%. Although the % RPD exceeded, the values were below the level of reporting (LOR). The exceeding of the % RPD is likely to be associated to the heterogeneity of the soil sample.

Two duplicate groundwater samples were collected for BH1 and BH7 and analysed for VOCs, SVOCs and metals. The % RPD were within the acceptable range for all the components that were analysed with the exception of chromium, where the % RPD was 78%. The reason for the exceeding of the % RPD is unknown.

Barium was detected in duplicate sample of BH1 at 59 μg/L, close to laboratory LOR of 50 μg/L, but not detected in sample BH1. This is likely due to the results of the sample concentration being too close to the laboratory LOR.

Laboratory QA Results

The laboratory performed an internal QA programme comprising method blanks, matrix spikes, surrogates, laboratory control samples and duplicates. The laboratory QA compliance assessment is summarized as below:

- Method blanks were performed and no concentrations were detected above the laboratory LOR.

- Surrogates were used on all gas chromatography (GC) analyses. All surrogate recoveries fall within the acceptable limits.

- Laboratory control samples analyses were performed and the results met acceptance recovery limits.

- Laboratory duplicates analyses were also performed and met the acceptance % RPD criteria.

- Matrix spikes were also performed and met the acceptance % RPD criteria.

Assessment Criteria - Dutch Intervention Value (DIV2000)

As soil and ground water standards have not been established in Malaysia, the analytical results were evaluated against the Dutch Guidelines (Appendix 8). The Dutch Reference Framework, which was issued by the Dutch Ministry of Housing, Physical Planning and the Environment in 2000, represents an internationally accepted evaluation framework. The DIV 2000 indicates the environmental quality level, above the maximum allowable risk of adverse effects on human and the environment is considered unacceptable.

It is recognized that the Dutch standards have been developed to assess the acceptability of contaminated soils and ground water at housing estates in the Netherlands, without reference to commercial or industrial land in Malaysia. Typically, the detection limits for many of the contaminants assessed under these standards are very low, especially for groundwater. On the basis, exceeding of the guidelines should not necessarily be interpreted as conclusive regarding the need for remediation of the site. If on the other hand concentrations of pollutants are below these criteria, then it would be reasonable to conclude levels were not of concern as Lijzen and et al (2001) had indicated in their evaluation of the RIVM report.

Also, the site being investigated is close to the coastal area, which is similar to the profile of Netherlands.

Field Observations

Visual and odour observation was conducted of the site and no chemical or organic odours were noted during soil sampling. The photo ionization detector (PID) was used to monitor the volatile organic compounds. In order to detect the VOCs in a PID, a gas sample is drawn into an ionization chamber where it is exposed to ultraviolet light. The released ions from compounds ionised generate an electrical signal from a collection electrode. The signal strength is related to the concentration of ionised compounds. The instrument measures those ionization potentials less than or equal to those of the lamp. Some explanation is provided in Appendix 3 on PID.

A PID screening of soil samples collected during the study indicated VOCs readings ranging from 5.0 ppm to 81 ppm (back ground PID readings ranging from 3.0 to 84 ppm). PID reading of each boring location is shown in the borehole log (Appendix 7).

Geology (subsurface soil)

The site is paved with asphalt in the exterior of the building area. Below the asphalt is a layer of fill material to a depth ranging from 0.2 meters bgl to 1.5 meters bgl. Underlying the fill layer are layers of clay and sand to a maximum depth of 5.0 meters bgl. The clayey sediments are of low to high plasticity and the sand is fine to coarse grained. This is by feeling the material during the drilling process when constructing the wells. Soil bore stratigraphy and notes taken during the construction of wells are shown in Appendix 7.

Hydrology

An elevation survey of the monitoring well was completed. Well levels at the top of the PVC casing were recorded. Measured groundwater levels are presented in Table 4.8 below. Water level for all the wells ranged from 0.434 to 1.222 metres. BH10 had the highest hydraulic water pressure located on the north west of the site and BH4 the lowest hydraulic water pressure located on the south east of the site.

Table 4.8: Groundwater Elevation to Determine Hydraulic Gradient

Monitoring Well ID	Top of PVC Casing	SWL form top of PVC Casing	SWL
	(RL in metres)	(metres)	(RL in metres)
Borehole BH1	10.109	1.142	8.967
Borehole BH2	10.077	0.959	9.119
Borehole BH3	10.006	0.472	9.534
Borehole BH4	9.988	1.210	8.778
Borehole BH5	10.108	0.907	9.201
Borehole BH6	10.008	0.666	9.342
Borehole BH7	10.005	0.758	9.247
Borehole BH8	10.135	0.903	9.232
Borehole BH9	10.166	0.855	9.311
Borehole BH10	10.000	0.434	9.566
Borehole BH11	10.000	0.642	9.358
Borehole BH12	10.656	1.172	9.482
Borehole BH13	10.588	1.321	9.267
Key: RL – denotes Reduced Level SWL – denotes Standing Water Level Temporary berchmark datum = 10.000 m			

Groundwater Elevation

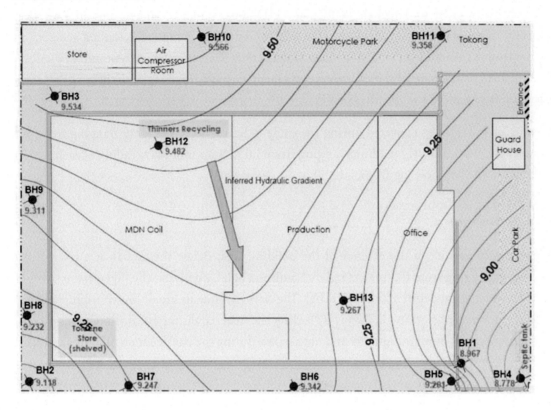

Figure 4.7: Layout Showing the Groundwater Elevation and Hydraulic Gradient.

Based on the survey and groundwater level gauging data, the inferred ground water flow direction is south, towards the coastline.

Groundwater elevations and the predicted hydraulic gradient are shown in the Figure 4.7 drawn and analysed on Surfer Six software. Some notes and explanation on the software programme is provided in Appendix 5.

Groundwater quality parameters were measured during the sampling event, after the monitoring wells had been purged. The results are summarised in Table 4.9.

Table 4.9: Data Collected during the Preparation and Development of Monitoring Wells for Soil and Groundwater Investigation.

Well ID	Water Level	V/P	Y/R	SL	Colour	Odour	Sheen/ Product	Temp	pH	EC	DO	Redox	Content
Borehole	mTOC	L						°C		µS/cm	mg/L	mV	
BH1	1.142	14	Very poor		Dark grey	Nil	Nil	27.9	7.08	31,200	0.25	90	PD
BH2	0.959	32	Good		Low brown	Nil	Nil	28.6	6.77	7,210	1.12	-41	
BH3	0.432	25.5	Poor		Med brown	Nil	Nil	27.2	6.31	9,740	0.01	-92	PD
BH4	1.210	15	Poor	High	Grey	Yes, organic	Nil	27.7	6.10	2,820	1.45	-6	PD
BH5	0.907	30	Good	High	Grey	Yes, organic	Nil	29.3	6.30	1,180	0.78	-92	
BH6	0.666	33	Good	High	Brown	Nil	Nil	30.8	6.30	22,820	650	-110	
BH7	0.758	33	Good	High	Grey	Yes, organic	Nil	31.3	6.27	1,682	1,180	-97	
BH8	0.903	High	Poor	High	Brown	Yes, organic	Nil	30.4	6.02	3,020	0.61	-42	PD
BH9	0.855	15	Poor	High	Grey	Nil	Nil	30.7	6.12	3,610	0.29	-76	PD
BH10	0.434	19	Poor	High	Brown	Yes, organic	Nil	31.7	6.07	3,460	0.80	-63	PD
BH11	0.642	20	Poor	High	Brown	Nil	Nil	33.2	5.92	2,289	1.06	-46	PD
BH12	1.174	29	Poor	High	Brown	Nil	Nil	27.5	5.93	3,290	2.09	-48	PD
BH13	1.321	14	Poor	High	Grey	Nil	Nil	27.7	5.90	6,200	0.62	-19	PD

Key:
Water level measured in metres from top of casing (mTOC)
SL – sedimentation load
Y?R – yield recovery
V?P – volume purged
PD – purged dry
Temp – temperature
EC – electrical conductivity
DO – dissolved oxygen
Redox – oxidation reduction potential

Temperature of groundwater ranged between 27.2 °C and 33.2 °C, one metre below ground level. The data indicates an even temperature throughout the plot of land being investigated. There are also no underground hot springs.

The pH of water ranged between 5.90 and 7.08, indicating a slight acidic to neutral condition. The value 7 denotes the neutral condition for pH measurements. As the plot of land is close to the shore line it is quite normal for the water to be slightly acidic in nature due to tide variation.

Electrical conductivity (EC) ranged between 1,682 µS/cm and 31,200 µS/cm, indicating a high volume of total dissolved solids in the ground water. The reason for the variation in EC across the site is due to the variation in soil permeability.

Dissolved oxygen (DO) values ranged between 0.01 mg/L and 2.09 mg/L, indicating a low level of DO in the ground water. Groundwater naturally contains low concentrations of oxygen. This is due to minimal re-aeration resulting from its laminar flow below the surface. Also, sites that have plants and vegetative growth will have increased DO present. This is due to plants during the photosynthesis process release oxygen into the water. Whereas, ninety five percent of the investigated site is covered with concrete and asphalt, as such it will have low levels of DO.

Oxidation reduction (redox) potential ranged between -110 mV and 90 mV, indicating slightly reducing to oxidising conditions. Redox is the term derived from *red*uction and *ox*idation. During the process there is transfer of electrons from one phase to another. Thus the condition applies for biodegradation to occur. As indicated in the USEPA (1997) biodegradation is the most important destructive natural attenuation mechanism. In most cases, degradation is a complex reduction or oxidation reaction process and can be conducted in a controlled state.

Soil and Groundwater Results

Soil Results

Soil sample results are summarised in Table 4.10 below.

Table 4.10: Summary of Results in the Investigation of Soil

Soil Results			VOC	sVOC	Metals							
					Ar	Cd	Cr	Pb	Se	Ag	Ba	Hg
Sample ID	Depth m (bgl)	Soil Type	mg/kg	mg/kg	mg/kg							
LOR					1	0.05	0.05	1	5	0.5	1	0.5
BH1	1.0	Fill	nd	nd	nd	nd	2.55	14	nd	nd	14	nd
BH2	0.5	Silty sand	nd	nd	nd	nd	1.75	35	nd	nd	10	nd
BH3	0.5	Gravelly sand	nd	nd	nd	nd	0.6	42	nd	nd	9	nd
Dup BH3	0.5		nd	nd	nd	nd	0.28	31	nd	nd	9	nd
BH4	1.7	Clay	nd	nd	nd	nd	11	25	nd	nd	12	nd
BH5	1.5	Sandy gravel	nd	nd	nd	nd	2	94	nd	nd	40	nd
Dup BH5	1.5		nd	nd	nd	nd	2	64	nd	nd	28	nd
BH6	1.5	Gravelly sand	nd	nd	nd	nd	2	45	nd	nd	15	nd
BH7	1.5	Sandy clay	nd	nd	nd	nd	3	36	nd	nd	16	nd
BH8	1.5	Gravelly clay	nd	nd	11	nd	2	1,020	nd	nd	914	nd
BH9	1.5	Sandy clay	nd	nd	nd	nd	3	576	nd	nd	880	nd
BH10	1.5	Clayey sand	nd	nd	nd	nd	1	39	nd	nd	22	nd
BH11	1.5	Clayey sand	nd	nd	nd	nd	2	29	nd	nd	19	nd
BH12	1.5	Clayey sand	nd	nd	nd	nd	3	60	nd	nd	22	nd
BH13	1.7	Gravelly sand	nd	nd	nd	nd	3	51	nd	nd	20	nd
Assessment Criteria DIV (2000)			NS	NS	55	12	380	530	100	15	625	10

Key:
LOR – Level of Reporting
nd – concentration below the laboratory LOR

VOC and SVOC were not detected in soil. With regards to metals in soil, lead and barium were above the DIV 2000 standards. Both the metals exceeded at two locations namely bore holes BH8 and BH9.

Barium Contamination in Soil

MiniTab was used to analysis the significance of lead contamination in the soil collected from the site.

From Table 4.11 below the following data and results were obtained.

Null hypothesis, H_0: Barium contamination in soil samples = 625 mg/kg

Alternative hypothesis, H_1: Barium contamination in soil samples < 625 mg/kg

P-value obtained is 0.000

Sample size is 13

The mean is 153.31 mg/kg

Standard deviation is 330.23 mg/kg

Table 4.11: Statistical Analysis of Barium in Soil – One-Sample T-test: Barium in Soil (mg/kg). Test for mu = 625 vs <625

One-sample T-test: Barium Contamination (mg/kg) Test for mu = 625 vs < 625							
Variable	N	Mean	St. Dev	SE Mean	95% Upper Bound	T	P
Barium (mg/kg)	13	153.31	330.23	91.59	316.54	-5.15	0.000

A total of 13 samples were analysed. The mean computed was 153.31 mg/kg and eleven of the points were below the mean. Two of the sampled points were above the mean and were at 914 and 880 mg/kg. Both the sampled points were located at the south west of the site and they can be considered as hot spots.

The null hypothesis was taken as 625 mg/kg. This is the acceptable limit by DIV (2000) Standard. Currently there are no Malaysia Standards.

Since the P-value is 0.000 and we are testing at a 5% level of significance, the test result shows as 'significant', thus we need to reject the null hypothesis in favour of the alternative hypothesis.

It can be concluded from the data above that with a CI of 95%, the probability of finding another hot spot is very low. Further investigation is not necessary and it can be deduced that there is barium contamination at the south west corner of the site. BH8 and BH9 are the major hot spots as can be seen from the graph in Figure 4.8 below. The two outliners on the right hand side of the graph are the hot spots with high content of barium.

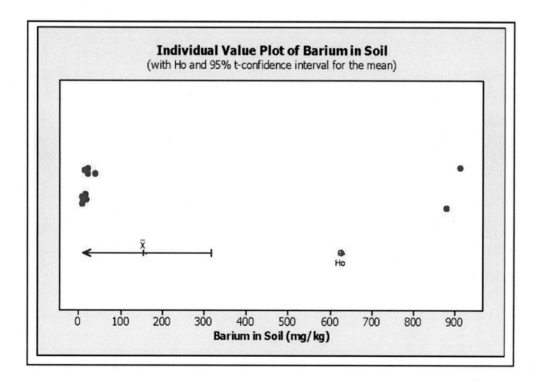

Figure 4.8: Graphical Analysis of Barium Concentration in Soil

Figure 4.9 is showing the spread of barium contamination on the site. Surfer Six software was used for the analysis and plotting the barium concentration distribution on the site. The two locations BH8 and BH9 are the hot spots located on the west side of the site. The spread is concentrated on the west boundary.

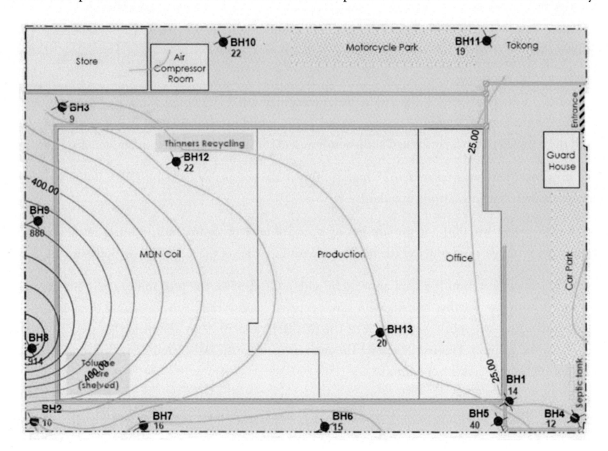

Figure 4.9: Layout of Site Indicating Concentration of Barium Contamination.

Lead Contamination in Soil

MiniTab was used to analysis the significance of lead contamination in the soil collected from the site. From Table 4.12 the following data and results were obtained.

Null hypothesis, H_0: Lead contamination in soil samples = 530 mg/kg

Alternative hypothesis, H_1: Lead contamination in soil samples < 530 mg/kg

P-value obtained is 0.000 (0.3%)

Sample size is 13

The mean is 158.92 mg/kg

Standard deviation is 298.38 mg/kg

Table 4.12: Statistical Analysis of Lead in Soil

One-sample T-test: Lead Contamination (mg/kg)						
Test for mu = 530 vs < 530						
Variable	N	Mean	St. Dev	SE Mean	95% Upper Bound	T	P
Lead (mg/kg)	13	158.92	298.38	82.76	306.42	-4.48	0.000

A total of 13 samples were analysed. The mean computed was 158.92 mg/kg and eleven of the points were below the mean. Two of the sampled points were above the mean and were at 1020 and 576 mg/kg. Both the sampled points were located at the south west of the site and they can be considered as hot spots.

The null hypothesis was taken as 530 mg/kg. This is the acceptable limit by DIV (2000) Standard. Currently there are no Malaysia Standards.

Since the P-value is 0.000 and we are testing at a 5% level of significance the test result shows as 'significant', thus we need to reject the null hypothesis in favour of the alternative hypothesis. It can be concluded from the data above that with a CI of 95%, the probability of finding another hot spot is very low. Further investigation is not necessary and it can be deduced that there is lead contamination at the south west corner of the site. BH8 is the out liner on the right hand side, which is the major hot spot as can be seen from the graph in Figure 4.10 below. Most of the other sample results are concentrated on the left side of the graph.

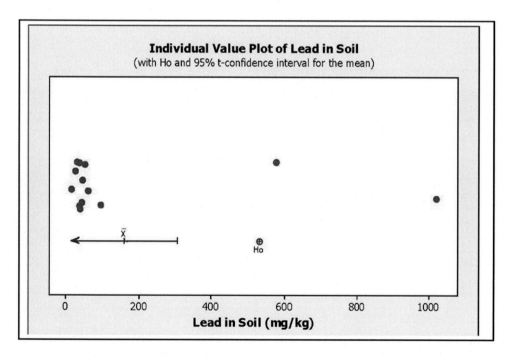

Figure 4.10: Graphical Analysis of Lead Concentration in Soil

Figure 4.11 is the layout showing all the lead levels on site obtained from the results in the investigation. Surfer Six was used to analysis and plot the lead concentration distribution on site. Lead concentration is mainly concentrated on the west side of the site. BH8 has the most concentration followed by BH9 and both the locations are very close to the boundary.

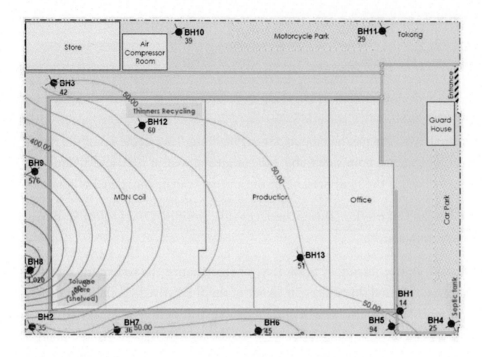

Figure 4.11: Layout of Site of Lead Contamination in Soil

Groundwater Results

Summary of groundwater results are shown in Table 4.13 below.

Table 4.13: Laboratory Results of Groundwater Investigation.

Groundwater Results		1, 1 – dicholoroethene	VOC	sVOC	Metals									
					Ar	Cd	Cr	Cr6	Pb	Se	Ag	Ba	Hg
Sample ID	Toulene µg/L	µg/L	µg/L	µg/L	µg/L									
LOR	5	5	-	-	50	1	1	50	10	100	10	50	0.2
BH1	nd	nd	nd	nd	nd	nd	14	nd	nd	100	nd	nd	nd
Dup BH1	nd	nd	nd	nd	nd	nd	32	nd	nd	nd	nd	59	nd
BH2	8	nd	8	nd	nd	nd	20	nd	nd	nd	nd	3.780	nd
BH3	nd	24	24	nd	nd	nd	3	nd	nd	nd	nd	131	nd
BH4	nd	nd	nd	nd	nd	nd	28	nd	nd	nd	nd	nd	nd
BH5	nd	nd	nd	nd	nd	nd	34	nd	nd	nd	nd	nd	nd
BH6	nd	nd	nd	nd	nd	nd	63	nd	nd	nd	nd	nd	nd
BH7	nd	nd	nd	nd	nd	nd	38	nd	nd	nd	nd	nd	nd
Dup BH7	nd	nd	nd	nd	nd	nd	35	nd	nd	nd	nd	nd	nd
BH8	nd	nd	nd	nd	nd	nd	18	nd	nd	nd	nd	1.420	nd
BH9	nd	nd	nd	nd	nd	nd	27	nd	nd	nd	nd	510	nd
BH10	nd	nd	nd	nd	nd	nd	39	nd	nd	nd	nd	206	nd
BH11	nd	nd	nd	nd	nd	nd	56	nd	nd	nd	nd	199	nd
BH12	nd	nd	nd	nd	nd	nd	23	nd	nd	nd	nd	173	nd
BH13	nd	nd	nd	nd	nd	nd	36	nd	nd	nd	nd	392	nd
Assessment Criteria DIV (2000)			NS	NS	55	12		380	530	100	15	625	10

Key:
LOR – Level of Reporting
nd – concentration below the laboratory LOR
NS – guidelines not specified

Toulene was detected at 8 μg/L at BH2 and was not detected at the balance of the boreholes. The DIV for toulene is 1000 μg/L, as such the detected value at BH2 is very low. 1, 1-decholoroethene was detected at 24 μg/L at BH3 and was not detected at the rest of the wells. The DIV for 1, 1-decholoroethene is 900 μg/L, as such the detected value is low at BH3.

Barium Concentration in Groundwater

MiniTab was used to analyse the significance of barium concentration in groundwater collected from the site. From Table 4.14 below the following data and results were obtained.

Null hypothesis, H_0: Barium concentration in Groundwater = 625 μg/L

Alternative hypothesis, H_1: Barium concentration in Groundwater < 625 μg/L

P-value obtained is 0.390

Sample size is 13

The mean is 542.77 μg/L

Standard deviation is 1042.03 μg/L

Table 4.14: Statistical Analysis of Barium in Groundwater.

One-sample T-test: Barium Contamination (μg/L) Test for mu = 625 vs < 625							
Variable	N	Mean	St. Dev	SE Mean	95% Upper Bound	T	P
Barium (μg/L)	13	542.77	1042.03	289.03	1057.86	-0.28	0.39

A total of 13 samples were analysed. The mean was computed as 542.77 μg/L and eleven of the points were below the mean. Two of the sampled points were above the mean and were at 3780 and 1420 μg/L. Both the sampled points are located on the right hand side of the box plot graph (Figure 4.12) and they can be considered as hot spots of the site.

The null hypothesis was taken as 625 mg/kg. This is the acceptable limit by DIV (2000) Standard. Currently there are no Malaysia Standards.

Since the P-value is 0.390 (is greater than 0.05) and since we are testing at a 5% level of significance, the test result shows as 'non-significant', thus we do not reject the null hypothesis.

As can be seen from the graph (Figure 4.13) barium concentration in groundwater is found at the south west corner of the site. BH8 and BH2 are the major hot spots as can be seen from the site layout.

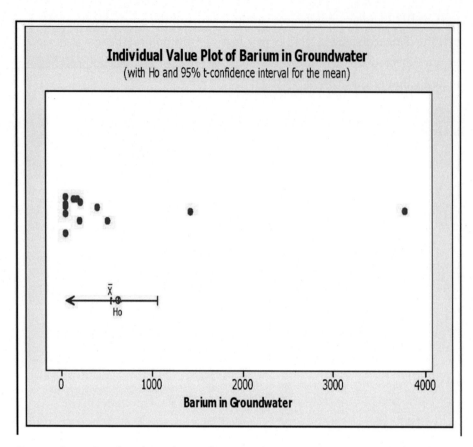

Figure 4.12: Graphical Analysis of Barium Concentration in Groundwater.

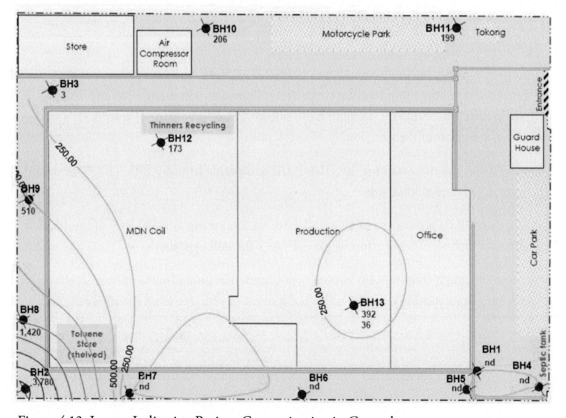

Figure 4.13: Layout Indicating Barium Contamination in Groundwater.

Chromium Concentration in Groundwater

MiniTab was used to analyse the significance of chromium contamination in the groundwater collected from the site. From Table 4.15 below the following data and results were obtained.

Null hypothesis, H_0: Chromium concentration in Groundwater = 30 µg/L

Alternative hypothesis, H_1: Chromium concentration in Groundwater < 30 µg/L

P-value obtained is 0.559

Sample size is 13

The mean is 30.69 µg/L

Standard deviation is 16.45 µg/L

Table 4.15: Statistical Analysis of Chromium Concentration in Groundwater

One-sample T-test: Chromium Contamination (µg/L) Test for mu = 30 vs < 30							
Variable	N	Mean	St. Dev	SE Mean	95% Upper Bound	T	P
Chromium (µg/L)	13	30.69	16.45	4.56	38.82	0.15	0.56

A total of 13 samples were analysed. The mean computed was 30.69 µg/L and seven of the points were below the mean. Six of the sampled points were above the mean and were at 34, 36, 38, 39, 56 and 36 µg/L. This can be observed in Figure 4.14; the distribution is evenly distributed on each side of the mean.

The null hypothesis was taken as 30 µg/L. This is the acceptable limit by DIV (2000) Standard. Currently there are no Malaysia Standards.

Since the P-value is 0.56 (is greater than 0.05) and since we are testing at a 5% level of significance, the test result shows as 'non-significant', thus we do not reject the null hypothesis. It can be concluded that there is presence of chromium concentration on the site. This can be observed graphically on Figure 4.14 and layout showing chromium concentration in groundwater on Figure 4.15.

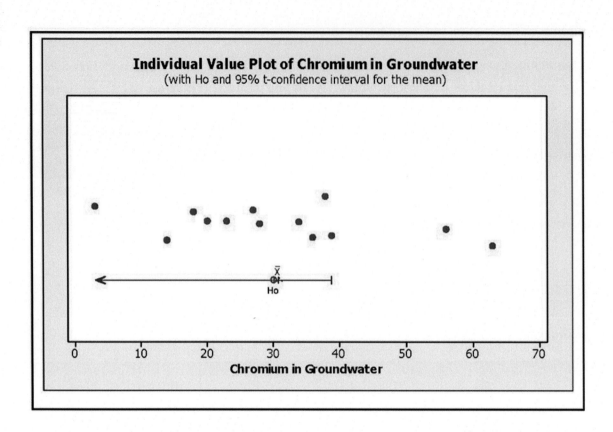

Figure 4.14: Graphical Indication of Chromium Concentration in Groundwater.

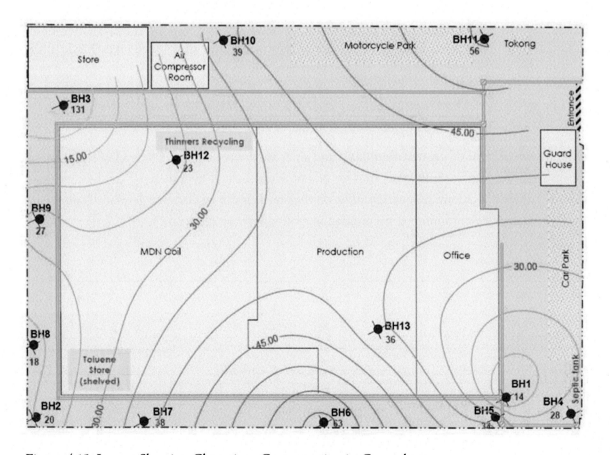

Figure 4.15: Layout Showing Chromium Concentration in Groundwater.

Previous Investigation

Three permanent monitoring wells BH1, BH2 and BH3 were installed during December 2004 investigation. Analytical results from soil and groundwater samples collected during this investigation are summarised as follows:

Metals analysis indicated chromium, lead and barium concentrations above the laboratory limit of reporting (LOR) but below their respective DIV (2000) assessment levels, chromium was 380 mg/kg, lead was 530 mg/kg and barium was 625 mg/kg in all soil samples collected at the Site.

Concentrations of dissolved metals were below the laboratory LOR in all groundwater samples with the exception of the following:

- Chromium in groundwater sample BH1 was 14 µg/L, BH2 was 20 µg/L and in BH3 was 3 µg/L. The concentrations were all below the DIV (2000) assessment level of 30 µg/L.

- Chromium in duplicate groundwater sample of BH1 was 32 µg/L and it is 2 µg/L above the concentration level of DIV (2000). The assessment level is 30 µg/L. Selenium in groundwater sample BH1 was 100 µg/L and the concentration level is below the DIV (2000) assessment level of 160 µg/L.

- Barium in groundwater sample BH2 is 3,780 µg/L and the concentration level is above the DIV (2000) assessment level of 625 µg/L.

- Barium in groundwater sample BH3 is 131 µg/L and the concentration level is below the DIV (2000) assessment level of 625 µg/L.

- VOC and SVOC analysis indicated concentrations level below the laboratory (LOR) in all soil and groundwater samples with the exception of:

- Toluene in groundwater sample BH2 is 8 µg/L and the concentration level below the DIV (2000) assessment level of 1,000 µg/L.

- 1, 1-DCE in groundwater sample BH3 is 24 µg/L and the concentration level below the DIV (2000) assessment level of 900 µg/L.

Off-site Assessment

Information obtained from interviews with a number of the HSE (Health, safety and Environmental) managers and staff at surrounding facilities is summarised as follows:

- The factory located to the northwest of the site which manufactures optical blanks for camera lenses and LCD lenses. Common chemicals used during the manufacturing process include thinners, isopropryl alcohol (IPA), powdered emery, polymer and lubricants. A wastewater treatment plant (WWTP) is present at the facility and caustic soda is used as part of the process waste water treatment. Sludge from the WWTP is stored on asphalt and bare ground, about 10 to 15 m from the site boundary. In addition, waste glass is stored about 5 m from the site boundary, adjacent to the southeast corner.

- The factory located two lots northeast of the site and serves as an assembly line for an electronics assembly multi-national. It was reported that chemicals used on-site include IPA (used for cleaning only), flux, solder wire and solder flux. The chemicals were reportedly stored on mosaic flooring. It was reported that there is no WWTP, aboveground or underground storage tanks at the facility.

- The factory located to the southeast of the site and which manufactures V-chips and E-chips (capacitors). The land has been leased since 2000. Common chemicals used at the facility include alcohol and electrolytes. There is no WWTP at the site.

- Another electronics factory located three lots northeast of the site currently trades in electronic components. It was reported that electronic components (termistor and varistor) were manufactured at the facility from July 1990 to April 2001. During this period thinners and alcohol were reportedly used on-site.

On-site Assessment

At the time of investigation, the site had been decommissioned by the previous owner. All information obtained relating to on-site activities is based on interviews with former staff and observations made by the auditors. Some documentation to cross reference information provided by the interviewees was verified and is summarised as follows: It was reported that in 1997, the site was purchased by the present owner, who operated the facility as a ceramic capacitor and inductor manufacturer.

In owner ceased its manufacturing activities in May of 2000 and the production area was leased to an electronics company who manufactured electronic boards using the surface mount technology. The manufacturer ceased its assembly operations in December 2003 and the facility has remained empty since then.

Chemicals used by the site were formerly stored in the store and production area on pallets. It was reported that there was no secondary containment available for chemicals. Chemicals stored and used on-site included alcohol, toluene, epoxy, flux, thinner, solder, hardener, colorant and silver paste.

Potential Sources of Metal Impact

Potential sources of metals identified on-site include:

- Lead (in the form of lead oxide) is commonly used in the glass industry to obtain a high refractive index for optical glasses.

- Barium (in the form of barium carbonate) is used to enhance the strength and scratch resistance in glass.

- Chromium is commonly found in the glass production to give glass an emerald green colour and ceramic capacitor manufacturing facilities (in the form of nickel -chromium) to produce high heat resistance ceramic capacitors.

Summary

Soil Results

Analytical results from soil samples collected during the investigation are summarised as follows:

- Concentrations of VOC were below the laboratory LOR in all samples analysed.

- Concentrations of SVOC were below the laboratory LOR in all samples analysed.

- Concentrations of metals were below the laboratory LOR in all samples analysed with the exception of:

  o Arsenic in soil sample BH8 was 11 mg/kg and concentration level is below the DIV (2000) assessment level of 55 mg/kg but above the USEPA Region 9 PRG of 1.6 mg/kg.

  o Total chromium in all samples ranged from 1 to 11 mg/kg. The concentrations levels are below both the DIV (2000) assessment level of 380 mg/kg and USEPA Region 9 PRG of 450 mg/kg.

Lead was detected.

Lead was detected at BH4 at a level of 25 mg/kg and the duplicate sample was 64 mg/kg. For BH5 it was 94 mg/kg, BH6 was 45 mg/kg, BH7 was 36 mg/kg, BH10 was 39 mg/kg, BH11 was 29 mg/kg, BH12 was 60 mg/kg and BH13 was 51 mg/kg. All the samples were below both the DIV (2000) assessment level of 530 mg/kg and USEPA Region 9 PRG of 750 mg/kg.

Sample from BH9 was 576 mg/kg and it was above the DIV (2000) assessment level of 530 mg/kg but below the USEPA Region 9 PRG of 750 mg/kg in, and Sample from BH8 was 1,020 mg/kg and it was above both the DIV (2000) assessment level of 530 mg/kg and USEPA Region 9 PRG of 750 mg/kg. Salatas et al (2004) had quoted in their study that according to USEPA's analysis database (USEPA 1997a), lead was the most frequent soil contaminant that exceeded screening criteria, followed by zinc, barium, cadmium, copper and beryllium. Of the industrial sites investigated, 25.9% had showed high content of lead. In another study conducted by Myers and Thorbjornsen (2004), lead was also one of the contaminant found on the site. The site was close to a catchment location, thus the site was classified as contaminated and was recommended to be remediated.

Barium was detected.

Barium concentration for BH4 was 12 mg/kg and the duplicate sample was 28 mg/kg, for BH5 it was 40 mg/kg, BH6 it was 15 mg/kg, BH7 it was 16 mg/kg, BH10 it was 16 mg/kg, BH11 it was 19 mg/kg, BH12 it was 22 mg/kg and for BH13 it was 20 mg/kg. All the above mentioned soil samples were below the DIV (2000) assessment level of 625 mg/kg and USEPA Region 9 PRG of 67,000 mg/kg. Whereas BH8 was 914 mg/kg and BH9 was 880 mg/kg, both these wells were above the DIV (2000) assessment level of 625 mg/kg but below the USEPA Region 9 PRG of 67,000 mg/kg. Study conducted by researchers (Turne and et al, 2006) for their post graduate work had showed high content of barium found in subsurface soil. The values ranged from 57.1 to 260.8 ppm from the Torrelles Municipal District in Catalonia, Spain.

Groundwater Results

Analytical results from groundwater samples collected during the Investigation are summarised as follows:

- Concentrations of VOC were below the laboratory LOR in all groundwater samples analysed.

- Concentrations of SVOC were below the laboratory LOR in all groundwater samples analysed.

- Concentrations of dissolved metals were below the laboratory LOR in all groundwater samples with the exception of total chromium.

- Total chromium concentrations were reported as follows:

- For well BH4 the groundwater concentration was 28 µg/L, for BH8 it was 18 µg/L, for BH9 it was 27 µg/L and BH12 it was 23 µg/L. All the results were below the DIV (2000) assessment level of 30 µg/L and the USEPA Region 9 PRG of 55,000 µg/L.

- For BH5 the concentration was 34 µg/L, BH6 was 63 µg/L, BH7 was 38 µg/L and the duplicate value was 35 µg/L, BH10 was 39 µg/L, BH11 was 56 µg/L and BH13 was 36 µg/L. All the above results were above the DIV (2000) assessment level of 30 µg/L, but below the USEPA Region 9 PRG of 55,000 µg/L.

Barium was detected at the following concentrations:

- The barium concentration for groundwater at BH8 was 1,420 µg/L. It is above the DIV (2000) assessment level of 625 µg/L but below the USEPA Region 9 assessment level of 2,600 µg/L.

- The barium concentration for BH9 was 510 µg/L, for BH10 it was 206 µg/L, BH11 it was 199 µg/L, BH12 it was 173 µg/L and BH13 it was 392 µg/L. All the concentration was below both the DIV (2000) assessment level of 625 µg/L and USEPA Region 9 assessment level of 2,600 µg/L.

Plotting of water level contours and contaminant contours for specific metals indicated that the source of contamination was isolated to the south western portion of the site. Also, the source of contamination was likely to have originated from off-site.

Chromium can be present in groundwater as a mixture of aqueous species with different charges (EPRI, 1984). The positive, neutral and negative charges on these species result in the distribution of Chromium on several different types of sorptive surfaces including clay and iron-oxide minerals. It was also indicated in the study that barium, lead and zinc tend to concentrate on clayey surfaces of soil.

According to the study conducted by Canter and Sabatini (1993) of 'Superfund' site contamination in United States, it had stated that multiple sources of contamination can effect groundwater supplies. Areas of concern was municipal landfills, industrial operations, leaking underground storage tanks, septic tank systems and dumping of uncontrolled hazardous waste. The main source of contamination was industrial operations and other activities associated with industries. Common metal contaminants found were chromium, lead and arsenic.

Quantitative Risk Assessment

Soil and Groundwater

Geology

The site is under lined by a layer of fill material to a depth ranging from 0.2 m to 1.5 m below ground level (bgl), beneath which are layers of clay and sand to a maximum depth of investigation of 4.0 metres bgl. The clay like sediments are of variable plasticity and the sand is fine to coarse grained.

Hydrogeology

Groundwater beneath the site is shallow and it is less than 1 metre bgl. The yield of the superficial aquifer is generally considered to be poor. Interpretation of water table elevations indicates that hydraulic gradient is to the south, toward the coast, which is located about 850 metres from the site. Field measurements indicate that groundwater beneath the investigated site is blackish is some areas, and as a result, it is unlikely to be utilised as a viable, potable source of water. No abstraction wells were identified on-site and also at surrounding properties. Movement of contaminants originating at the surface is controlled by the flow path according to Aiyeasanmi et al (2004). The study indicated that under a given hydraulic gradient the more permeable the earth materials, the more will be the velocity of contaminants flow along the flow path.

Sources of Contamination

In relation to the initial stage of this assessment, contaminants of potential concern are those compounds detected in soil or groundwater at concentrations in excess of the relevant screening criteria. Initially, Tier 1 screening criteria have been selected to be protective of human health. A range of values from the Malaysian Drinking Water Standards, DIV (2000) assessment levels and the US EPA Region 9 PRGs have been provided for comparative purposes.

Contaminants of potential concern identified at the site are summarised as follows:

- Soil - Lead, barium and arsenic were detected at concentrations in excess of at least one of the assessment criteria, and

- Groundwater - Total chromium and barium were detected in groundwater at concentrations close to or exceeding one or more assessment criteria.

Pathways

For the exposure pathway of a given source of chemical contamination to be considered complete, and hence creating a risk to a specified receptor, the following elements are required:

- A release mechanism - it can be a spill, through leakage, by wind erosion, infiltration on runoffs.

- A retention or transport medium - by air, soil or groundwater.

- An intake mechanism - by inhalation, ingestion or dermal contact.

Exposure pathways may be either direct or indirect. Direct exposure paths will be through soil ingestion or dermal contact. Indirect exposure occurs at a location or medium removed from the original source and includes indoor air quality. The majority of the site's surface is covered, either with concrete foundation or asphalt, therefore direct exposure to any impacted soil would only likely occur during;

- Periods of excavation, or

- If soil was brought to the surface during any piling activities.

- As the groundwater below the site currently does not represent a potable source of water and no shallow abstraction wells were observed or reported in the area, the only exposure scenario to impact groundwater is the same as that for soil.

- Exposure as a result of inhalation of contaminated vapours is not an issue at this site, as volatile compounds are not present at the site above preliminary screening levels. As the site is completely covered, potential exposure as a result of inhalation of contaminated dust particles would only be likely during any excavation or piling work, and is considered a low risk for this site.

Based on the investigative sample results and analysis both the soil and groundwater exposure mechanisms represent short-term scenarios, no chronic exposure scenarios are recognised for the site.

Receptors

Potential on-site human receptors include:

- Employees working on the site including the guards,

- Site visitors, and

- Construction workers.

Site employees are considered to have the greatest potential for chronic or long -term exposure to on-site contamination. However, the nature of the metals contamination detected on the site in both soil and groundwater, coupled with the hard standing site coverage, indicates that the risk of such exposure actually occurring may be limited under normal operating conditions.

It is considered that the potential for exposure of transient site visitors, such as customers, to the metals detected in soil and groundwater would be less than that of site employees. Accordingly, exposure pathways to these receptors are not deemed to be significant under current site conditions. Whilst not significant under normal site operating conditions, exposure pathways to on-site construction workers could become relevant in the event that works are undertaken which would facilitate access to soil and/ or groundwater. Construction workers on adjoining properties are considered to have only transient potential exposure to the metals detected in soil and groundwater, and as a result, exposure pathways to these receptors are not deemed to be significant.

Quantitative Risk Assessment

Although a number of metals such as arsenic, lead and barium in soil and total chromium and barium in groundwater exceeded preliminary screening levels adopted for this investigation, completion of a quantitative risk assessment indicated the following:

- Barium in soil exceeded the DIV (2000) assessment level of 625 mg/kg, but not the USEPA Region 9 PRG of 67,000 mg/kg at two locations (914 and 880 mg/kg). It is important to note that the DIV (2000) assessment level is based around protection of soil ecosystems and human health, two numbers are derived by RIVM (2001) and the smaller is chosen to represent the DIV assessment level. In the case of barium, the DIV (2000) assessment level of 625 mg/kg is the ecological protection number designed to protect soil invertebrates, the human health value is 9,340 mg/kg as indicated in RIVM (2001). The human health related screening values are significantly higher than any detected value on-site, indicating that potential health risks are not an issue. A new ecological value has been proposed for barium of 890mg/kg, for which there was only one sample that exceeded at the investigated site.

- In addition, statistical analysis was completed to identify the 95% upper confidence limits (95% UCL) of the measured mean and to compare the upper 95th percentile with the soil screening criteria as recommended in (CLR 7, 2002). The barium 95%UCL has been calculated as 276 mg/kg. This result is significantly below both the ecological and health based elements of the DIV (2000) and therefore, barium in soil is not considered to be present as a significant contaminant of potential concern.

- Lead in soil exceeded the DIV (2000) assessment level of 530 mg/kg at two sampling locations (576 and 1,020 mg/kg) and the USEPA Region 9 PRG assessment level (750 mg/kg) at one location. As for barium, statistical analysis of the 15 lead data results was undertaken. The 95% UCL of the arithmetic mean was calculated as 271 mg/kg. This result is below both the DIV (2000) and US EPA assessment levels and therefore, lead in soil is not considered to be present as a significant contaminant of potential concern.

- Arsenic in soil was detected in soil above the laboratory LOR in one sample at 11 mg/kg. This did not exceed the DIV (2000) of 55 mg/kg but did exceed the USEPA PRG of 1.6 mg/kg. The PRG is based on highly conservative IRIS database Slope Factors for the carcinogenicity of arsenic. Extensive review of this work has been carried out by RIVM (2001) and they have recommended a generic health based risk limit for soil for arsenic of 576 mg/kg. This value is in keeping with the Australian Industrial Health Investigation Level (HIL) of 500 mg/kg and the UK Industrial Soil Guideline Value of 500 mg/kg.

- Clearly there is a major discrepancy between the USEPA and other approaches. As discussed above the PRG is a preliminary screening number based on a rigid hierarchy of toxicity data. The DIV is based on an independent review of all the available toxicity data for arsenic and after careful consideration of this data RIVM concluded 'There were suggestions that inorganic arsenic is a human carcinogen, but it was concluded that available studies demonstrate too few arguments to propose an exposure limit value by means of a non - threshold extrapolation. Carcinogenic effects can only be observed when also toxic effects are noticed'. Therefore, arsenic in soil is not considered to be present as a significant contaminant of potential concern.

- Total chromium in groundwater (Figure 4.14 shows the data points graphically) exceeded the DIV (2000) assessment level of 30 µg/L in nine samples (32, 34 (x2), 35, 36, 38, 39, 56 and 63 µg/L) and in two samples of the Malaysian Drinking Water Standards (2006) of 50 µg/L. The Malaysian Drinking Water standard is based on the WHO guideline value of the same value. The WHO guideline value is based primarily on the considerably more toxic/carcinogenic hexavalent chromium. The WHO guideline states that, 'In principle, it was considered that different guideline values for chromium (III) and chromium (VI) should be derived'. However, current analytical methods favours the WHO (1996) guideline values for total chromium.

- Analytical results from this site have shown that the chromium is present in the trivalent oxidation state. The analytical results for hexavalent chromium were below the laboratory LOR for all groundwater samples collected during the investigation. A drinking water standard based on total chromium is therefore less valid in assessing the significance of the chromium identified at the site. It is well established knowledge that chromium is significantly less toxic and carcinogenic in its trivalent than its hexavalent form. For non-carcinogenic chemicals, it is assumed that a dose threshold, below which no adverse effects will occur, exists. This threshold dose level is termed the Tolerable Daily Intake (TDIs), and can be obtained from the US EPA IRIS Database, which is updated periodically. The TDI for trivalent chromium provided by IRIS is 1.5mg/kg per day (last reviewed in March 1998). Using this value in the formula for determination of the TDI, a drinking water guideline value of 4,500 µg/L can be derived for trivalent chromium.

- Clearly the detected values of trivalent chromium on-site are significantly below this value and are not therefore considered to represent a potential risk to human health. Trivalent chromium was not detected above the Dutch ecological criterion of 220 µg/L as referenced in the RIVM (2001), and is not therefore considered to represent a potentially significant risk to ecology. Therefore, total chromium in groundwater is not considered to be present as a significant contaminant of potential concern.

- Barium in groundwater exceeded the DIV (2000) assessment level of 625 µg/L and Malaysian Drinking Water Standard of 1,000 µg/L in two samples (1,420 and 3,780 µg/L). The Dutch ecological criterion of 7,100 µg/L for barium and the USEPA Region 9 tap water PRG of 2,600 µg/L was not exceeded by any sample.

- The significance of this sample with regard to human health has to be taken in context with the conceptual site model as a whole. It is apparent from studies of the site that groundwater is not used for drinking water in the vicinity of the site and as such the detected concentration of barium does not pose a significant risk. However, should groundwater beneath the site be used as a potable water resource (which is considered unlikely), groundwater quality beneath the site would need to be considered in more detail. Therefore, under the current land use scenario, barium in groundwater is not considered to be present as a significant contaminant of potential concern.

As the site is predominantly covered with hard standing and buildings, restricting potential direct exposure pathways to soils and groundwater, and as the identified contaminants are not volatile therefore limiting indirect exposure to vapours, and as no groundwater abstraction points were found to be located in the vicinity of the site no significant pollutant linkages between the soil and groundwater contamination and the site users could be identified. Should any future excavation works be undertaken at the site, it is recommended that EH&S measures are implemented to minimise any potential exposure risks as a result of impacted soil and/or groundwater being brought to surface.

Lead Paint Analysis

Eight paint samples were collected for lead content analysis. Results indicated that one sample, collected from the stairway handrail leading to the 1st Floor canteen, had a lead concentration in excess of the other seven samples. There are currently no available standards as 'screening numbers' of any lead paint for comparison of measurements to determine their potential impact on health. However, there is a US and an Australian standard for maximum allowable concentration of lead in domestic paint. The Australian number is 0.1% when dried, but this standard is only a quality standard for paint production and is not necessarily 'health-based' in the classical risk assessment (dose response) context. The other closest standard is the heavy elements requirements for children products, such as the EN71, ISO 8124 and ASTM F963 which calls for the total lead content to be below 600 ppm (parts per million) for surface coated items. This standard will only be applicable if the surface coating becomes loose and is absorbed or inhaled in our digestive or respiratory system. Although ingestion by young children is generally considered to be the exposure pathway of greatest concern for lead based paint, ingestion of lead base paint is not likely to occur at the site due to the industrial nature of the site and the fact that young children are not allowed on the premises. As such this standard is not applicable but can only be used as a guideline. All samples, other than the one collected from the stairwell were below this standard and, as a result, were not considered in any potential risk related scenarios. There is no straightforward technique to provide a realistic model for the health risk associated with regard to exposed skin coming into contact with the handrail at the site. If a conservative type approach is considered then a pathway exists by which humans could come into contact with the paint, although exposure is likely to be very low and we do not know the exposure point concentration" which is the concentration of lead that is absorbed through the skin. The analysis result is of paint chips, which will give a worst case scenario rather than a swab test, which would provide a more realistic indication of paint transfer during contact. Also, the area of lead based paint at the site is presently in good condition, inhalation does not appear to be an exposure pathway under the existing conditions. Taking all of the above into consideration, it can be considered that the potential health risk from the presence of lead in paint on the handrail is likely to be negligible. However, there is a small degree of uncertainty associated with this risk.

Suspected Asbestos Containing Material Sample Analysis

The samples were analysed using the polarised light microscopy (PLM) technique was used and EPA PLM protocol for determining if asbestos fibres exist. Bulk samples were not collected from homogenous areas where the assessor was sufficiently satisfied that the material was fibreglass, foam glass or other easily identifiable non ACM. Analysis of the single material sample collected from the site indicated that it was not asbestos containing material.

Phase Three Study - Remediation

Results for asbestos, soil, groundwater and lead in paint are discussed in the earlier section of the chapter. There was no contamination detected for asbestos on the site. As such no remediation is necessary.

For soil and groundwater although there were some hot spots identified. The hot spots were assessed applying the risk assessment model. The radox level of the site ranged from -110 mV to 90 mV. This indicates that the site as slightly reducing to oxidising condition. This indicates that the site condition allows for biodegradation to occur. Biodegradation is the most important destructive natural attenuation. The site is found to be safe for the site occupants. Thus no remediation is required for soil and groundwater.

For lead in paint, there was one location that had an elevated (orange paint) reading. (The seller had removed the hand railing and disposed it off as hazardous waste to Kualit Alam who is the registered waste collector, prior to the sale completion). All other location paints are within the limits and would not cause any ill effect to occupants. As such the site is safe and no further remediation for paints on the site is required.

CHAPTER FIVE

CONCLUSION AND RECOMMENDATION

Conclusion

The Federal Cabinet in Germany has approved a draft law (ENDS, 2006) in place for the individual operators of the sites in preventing and remediating environmental damage. The new law will implement the 2004 EU environmental liability directive, which allows the operator of the site to be sued if owing to their activities they damage land, water, protected habitats and species. This draft legislation is pending the approval of the lower house before it will take effect.

More and more governments are taking this move to protect the soil and groundwater. The Asian countries are also venturing into specifying and controlling the contamination of soil and groundwater. Taiwan has similar laws but the requirements are not as extensive as the US. It also specifies limited number of chemical and metal contaminants. China is currently (2006/7) drawing up such requirements as well. It is believed that within two years the requirements (regulations) will become enforceable. Malaysia on the other hand will also start drawing up such requirements in time to come. But nevertheless many multi-national corporations have started practising and are taking preventative measures to avoid future liabilities. So when a new site is acquired the due protocols need be adhered to. Since there are no protocols in Malaysia the corporation will use internationally recognized or their home base protocols.

The audit manual/protocol (Table 5.1) applied in this study was able to identify issues located on the site. Based on this study, the manual can be easily applied by an environmentalist with basic knowledge. Its content has been arranged in a systematic way for ease of comprehension and would be an acceptable approach for future investigations. It also fulfils the need of the topic of this thesis for the protocol development. It is more comprehensive than any individual standard. The local Department of Environment has accepted the results from this study as a base line for the investigated site. Inhalation of lead paint dust during maintenance activities or some construction on the handrail may result in a potential health risk to workers. Therefore, it is recommended that the purchaser develops and implements a lead based paint management programme to minimise the potential health risks posed by the lead based paint. The programme should include employee notification of the location of lead based paint, training on the hazards, associated health risk and effects and training on the proper handling of lead based paint. In the event that the lead based paint is disturbed or found to be severely deteriorated than provision should be in place for exposure monitoring of site workers, as well as possible biological and health monitoring if exposure monitoring shows significant lead exposures. The lead based paint management programme should also include regular inspection and maintenance of lead based areas to ensure minimal disturbance or deterioration of existing surface conditions. (To mitigate the risk, the seller removed the handrail and disposed it as scheduled waste prior to the sale transaction. The site is deemed to be free from any further risk).

The Malaysian Department of Occupational Safety and Health (DOSH) regulates exposures to airborne asbestos fibres under the Occupational Safety and Health (USE and Standards of Exposures of Chemicals Hazardous to Health Regulation 2000). The exposure to all forms of asbestos (except crocidolite) is limited to 0.1 fibres per millilitre of air on an 8-hour Time Weighted Average (TWA) basis. There are however, no specific regulation in Malaysia related to the presence or amount of ACM in buildings or in equipment. It should be noted that ACM with more than 1% asbestos when disturbed has the potential of causing airborne concentration at or about the permissible exposure limit of 0.1 fibres per millilitre of air. Thus, the US EPA has limited the asbestos content of some material to 1%. However, asbestos has not been eliminated from all manufactured building materials. Therefore, the age of a building or remodelling project cannot be the basis for assuming that a material does not contain asbestos. The EPA's clearance criteria are based on statistical comparisons of airborne asbestos concentrations inside a work area against those outside the work area. Phrase Contrast Microscopy (PCM) is an acceptable analytical method; the EPA has recommended 0.01 fibres per cubic centimetre as a limit for airborne fibre concentrations. In addition to the exposure criteria, the US EPA National Emissions Standard for Hazardous Air Pollutants (NESHAP) (40CFR, Part 61, subparts A and M) need to be followed during any maintenance or renovations activities which might disturb the ACM.

Based on the results of the research and investigation, the groundwater and soil contamination at the site was found not to pose a human health risk, the off-site source may remain for some time to come, causing an ongoing contamination issue at the site. The buyer may wish to develop a groundwater and soil monitoring programme in key wells to assess any potential changes in the groundwater and soil impact over time. The safety department of the company will need to develop procedures and documents identifying and specifying the area of contaminants. Any future construction or excavation work done on the site, safety precautions need to be administrated. Construction workers will be required to use the appropriate personal protective equipment when working on site. Groundwater if discharged during the excavation works need to be identified and stored appropriately. The owner will need to analyse the groundwater prior to disposal. If the groundwater is contaminated it needs to be treated prior to discharge or it can be sent to Kualiti Alam for disposal. Similarly, for the excavated soil, it has to be analysed. If contamination is identified than the soil needs to be disposed appropriately.

The site is safe to be used and no further remediation is required.

Table 5.1: Audit Protocol/Manual

Audit Protocol/Manual	
Flow	**Activities**
Phase 1	Purpose and Limitation Site Description Record Review Agency Contact and Investigation Previous Studies if Available Property Description Hazardous Material Drums and Storage Containers Solid Waste Water Supply and Waste Water Discharges Potential PCB Containing Equipment Underground Storage Distressed Vegetation Conditions of Surface Soil and Soil Disturbances Air Emissions and Odours
	Drilling (3 wells)
Phase 2	Asbestos Lead Based Paint Investigations Surrounding property Reconnaissance Hydrogeology Hydrology Drilling Soil Sampling Groundwater Sampling Project Cost Quality Assurance and Quality Control (QA/QC) Procedures
Phase 3	Remediation

Recommendation

Based on the research and investigation on the site under study, it is safe for the buyer to purchase the site. The liabilities identified are limited and does not pose a major concern for the buyer. Nevertheless, it is recommended that the buyer administers the following recommendations:

- To contact the national electricity board authorities and request to verify the oil in the transformer belonging to them which is located on the investigated site that it does not contain any pcb (polychlorinated biphenyls).

- Due to the small degree of uncertainty associated with the potential risk associated with the presence of lead in paints, the most cost effective action may be to have a procedure in place to use only paints free of lead for all future surface coatings.

The data was shared with the Department of Environment, Penang to establish the transparency of the investigative findings and also generate knowledge sharing between the government sector and MNCs.

REFERENCE

1. Aiyessanmi A.F., Ipinmoroti K.O. and Adeeyinwo C.E., (2004). *Baseline Geochemical Characteristics of Groundwater within Okitipupa South -East Belt of the Bituminous Sands Field of Negeria.* International Journal of Environmental Studies, 61: 1, 49-57.

2. ASTM E1527 (2000). American Society for Testing and Materials (ASTM) Standard Practice E1527, *"Standard Practice for Environmental Site Assessments: Phase 1 Environmental Site Assessment Process".*

3. ASTM E1528 (2000). American Society for Testing and Materials (ASTM) Standard Practice E1528, *"Environmental Site Assessment: Transaction Screen Process".*

4. ASTM F963 (2008). Standard Consumer Safety Specification for Toy Safety.

5. Canter L.W. and Sabatini D.A. (1994). *Contamination of public ground water supplies of Superfund sites.* International Journal of Environmental Studies, 46: 35-57.

6. CERCLA (1980). Comprehensive Environment Response, Compensation and Liability Act (Superfund), (1980). 42 USC § 9601 et seq., 1980.

7. Chapelle F.H. (1993). *Groundwater Microbiology and Geochemistry.* New York: Wiley.

8. CLR 4, (DOE, 1994). *Sampling Strategies for Contaminated Lands.* Contaminated Land Report 7.

9. CLR 7, (2002). *Assessment of Risks to Human Health from Land Contamination: an overview of the Development of Soil Guideline Values and Related Research.* Contaminated Land Report 7. Undated by Prof Paul Bardos 27/09/06.

10. Contaminated Land Management Act (1997). New South Wales, Australia.

11. Deutsch, W.J. (1997). *Groundwater Geochemistry: Fundamentals and Applications to contamination.* Lewis Publications, Boca Raton, Florida.

12. Diamond, S. (1985). *"The Bhopal Disaster: How it Happened",* The New York Times, January 28th.

13. DIV (2000). Dutch Ministry of Housing, Spatial Planning and the Environmental (MVROM), 2000. Circular in Target Values and Intervention Values for Soil Remediation. Directorate-General for Environmental Protection.

14. EN 71 (20030. Toy Safety Standards.

15. ENDS (2006). Europe Daily, 21 September 2006. http://www.endseuropedaily.com

16. Environmental Quality Act and Regulation (1974). Environmental Quality (Scheduled Wastes) Regulation 1989.

17. EPRI (Electric Power Research Institute). (1984). *Chemical Attenuation Rates, Coefficients, and Constants in Leachate Migration,* Volume 1: A Critical Review. Electric Power Institute: Plo Alto, California. EPRI EA-3356.

18. Feagin, J., Orum, A., & Sjoberg, G. (1991). *A Case for Case Study.* Chapel Hill, HC: University of North Carolina Press.

19. Finkel A. J. (1983). *Lead.* In Hamilton and Handy's Industrial Toxicology, Edition 4, Littleton, Mass, PSG Publishing.

20. Galliers Robert (1992). *Information Systems Research: Issues, Methods and Practical Guidelines,* Blackwell Scientific. ISBN-13: 978-0632028641

21. Garman & Clayton (1997). *Qualitative Research for the Information Professional: A Practical Handbook.* Library Association.

22. Geological Map of Peninsular Malaysia (1985). 8th Edition (scale 1:500,000), published by the Director-General of Geological Survey, Malaysia.

23. ISO 8124 (1997). Safety of toys - Part 3: Migration of Certain Elements.

24. ISO 14015 (2004). Soil Quality - Characterization of soil related to groundwater protection.

25. Kalelkar, A.S., (1988). *"Investigation of Large Magnitude Incidents: Bhopal as a Case Study,"* presented at the Institute of Chemical Engineers Conference on Preventing Major Chemical Accidents, London, May 1988.

26. Lijzen, J.P.A., Baars, A. J., Otte, P.F., Rikken, M.G.J., Swartjes, F.A., Verbruggen, A.P., Wezel, van. (2001). *Technical Evaluation of the Intervention Values for Soil/Sediment and Groundwater.* Dutch Institute of Public Health and the Environment (RIVM) Report No. 711701023.

27. Mackenzie L.D. and Cornell D.A. (1991). *Introduction to Environmental Engineering.* International edition, McGraw Hill, ISBN 0 -07-100828-4.

28. Malaysian Drinking Water Standard (2006). Twenty-fifth schedule, Regulation 294 (1) and 360B (3) of the Food Act 1983 (Act 281) as of 1st March 2006 and Regulations. International Law Book Services.

29. Myers J and Thorbjornsen, K. (2004). *Identifying Metals Contaminated in Soil: A Geochemical Approach.* Soil and Sediment Contamination, volume 13: 1 -16, 2004. ISSN: 1058-8337.

30. New Jersey's Environmental Clean-up Responsibility Act of 1984 (ECRA).

31. Plummer, L.N., Prestemon, E. C., Parkhurst, D. L. (1994). *An Interactive Code for Modelling NET Geochemical Reactions along a Flow Path.* U.S. Geological Survey Water resources Investigation Report 94-4169. 130 pp.

32. Public Health of Scotland Act (1897).

33. Resources Conservation and Recovery Act (1976). Later amended in 1984.

34. RIVM (2001). The National Institute of Public Health and Housing. RIVM report 128507010 - S.C. de Greeff, J.F.P. Schellekens, F.R. Mooi and H.E. de Melker.

35. Salatas, J. H., Lowney, Y. W., Pastorok, R. A., Nelson, R. R., and Ruby, M. V. (2004). *Metals tha Drive Health-Based Remedial Decision for Soils at US Department of Defence Sites.* Human and Ecological Risk Assessment Volume 10, No. 6, 2004: 983-997. ISSN 1080-7039.

36. SARA (1996). Superfund Amendment and Reauthorization Act.

37. Sax, N. I. and Lewis R.J., (1989). *Dangerous Properties of Industrial Materials,* 7th Edition, Van Nostrand Reinhold, New York.

38. Scottish Executive (2000). Contaminated Land (Scotland) Regulations 2000 (SI 2000/178). http://www.hmso.gov.uk

39. Selikoff, I. (1991). Mount Sinai School of Medicine, New York.

40. Shrivastava, Paul. (1987). *Bhopal - Anatomy of a Crisis*, Ballinger Publishing Co., Cambridge, Massachusetts.

41. Six Sigma Academy. (2002). The Black Belt Memory Jogger. A pocket guide for Six Sigma Success. 1st Edition. ISBN 1-57681-043-7

42. Smith DC, Spivack AJ, Fisk MR, Seifert R, Michaelis W. (2000). *Tracer-based estimates of drilling-induced microbial contamination of deep-sea crust.* Geomicrobiol J 17: 207-219.

43. Stake, R. (1995). *The Art of Case Research.* Newbury park, CA: Sage publication. ISBN 0-8039-5767-X

44. Stubbs, M.C. and Tang, O.H. (2004). *Assessing the Significance of Subsurface Vapour Migration into Indoor Air: A case study.* Brownfield Asia 2004, KL, 8-10 June.

45. Tellis, W. (1997, July). *Introduction to Case Study. The qualitative Report.* On line serial http://www.nova.edu/ssss/QR/QR3-2/tellis1.htm

46. Toxic Substances and Disease Registry (1988). Department of Health and Human Services Agency.

47. Turne, P., Bech, J., Longan, L., Turne, L., Reverter, R. and Sepulveda, B. (2006). *Baseline Concentrations of Potentially Toxic Elements in Natural Surface Soils in Torrelles (Spain)*. Environmental Forensics, Volume 7: 369-375, 2006. ISSN: 1527-5922.

48. USEPA TCLP (1990). US EPA: *"Hazardous Waste Management System; Identification and Listing of Hazardous Waste; Toxicity Characteristic Revisions."* Final Rule, Federal Register, Volume 55.

49. USEPA (1997). *Use of Monitored Natural Attenuation at Superfund, RCRA Corrective Action, and Underground Storage Tank Sites,* Office of Solid Waste and Emergency Response, Directive 9200.4-17, Washington, D.C. 1997.

50. USEPA (US Environmental Protection Agency) (1986). *Title 11- Asbestos Hazard Emergency Response.* Public Law 99-519, October 22, 1986.

51. USEPA (US Environmental Protection Agency). (1997a). *Cleaning up the Nation's Waste Sites: Markets and Technology Trends,* 1996 Edition. PB96-178041. Office of Solid Waste and Emergency Response, Washington, DC, USA.

52. United States Department of Health and Services, *"Toxicological Profile for Lead"*. The anaemia of lead poisoning: a review, British Industry Medical Journal, 23:83.

53. United States Department of Health and Human Services (1988). Agency for Toxic Substances and Disease Registry: *The nature and Extent of lead poisoning in children in the United States:* a report for congress.

54. Waldron H.A. (1966). *The anaemia of lead poisoning: a review*, British Industry Medical Journal, 23:83.

55. Walter, M.R. (1982). *Ground Water and the Rural Homeowner,* Pamphlet, U.S. Geological Survey, 1982.

56. WHO (1996). /Chromium in Drinking Water. Background document for development of WHO Guidelines for Drinking Water Quality, 2nd Edition, Volume 2. (WHO/SDE/WSH/03.04/04) World Health Organisation, Geneva.

57. Yin, R. (1993). *Application of Case Study Research.* Newbury Park, CA: Sage publication. ISBN 0-7619-2550-3

58. Yin, R. (1994). *Case Study Research: Design and Methods.* 2nd Edition. Thousand Oaks, CA: Sage publication.

59. 40 CFR 262.24 Federal Register. *Hazardous Waste Management and Characterising of Hazardous Waste.*

APPENDIX ONE

STATISTICAL ANALYSIS

Statistic - The Hypotheses

Researchers conduct a study to conduct a 'hunch'.

The hunch gives rise to a research hypothesis which the researchers try to establish as being true. Unfortunately, a decision in a hypothesis can never conclusively be defined as a correct decision. All the hypothesis test can do is minimise the risk of making a wrong decision.

Hypothesis tests involve 2 competing statements called the null hypothesis and the alternative hypothesis.

Null Hypothesis, H_0 - we test the null hypothesis. We determine how much evidence we have against H_0. The null hypothesis usually takes a sceptical view, there is nothing new or interesting or there is no effect.

Alternative Hypothesis, H_1 - the alternative hypothesis corresponds to the research hypothesis. It usually takes the form that something is happening, there is a difference or an effect, or there is a relationship. In most situations the researcher hopes to give support to H_1 by showing that H_0 is not believable.

Example in the case of lead contamination in paints, the hypothesis will be as follows;

Null Hypothesis, $\quad\quad\quad\quad H_0$: Lead Contamination in paint = 600 mg/kg

Alternative Hypothesis, $\quad\quad H_1$: Lead Contamination in paint < 600 mg/kg

The t-Test Statistic

This t-Test compares one sample average to a historical average or target. If the hypothesised value for the parameter is correct, then the estimate should not be too far from the hypothesised value. When the hypothesised value is correct, we assume the sampling distribution of the test statistic to be approximately a student distribution.

The P-value

The P-value is the conditional probability of observing a test statistic as extreme as that observed or more so, given that the null hypothesis is true.

The P-value for a t-Test

The P-value for a t-Test is the probability of getting an estimate as far away from the hypothesised value as our estimate, when H_0 is true. The P-value is compared to the decision criteria (α risk) and determined whether to reject H_0 in favour of H_1, or not to reject H_0. The criteria used as is as follow;

If the P-value is less than the α-risk, than reject H_0 in favour of H_1.

If the P-value is greater than the α-risk, there is not enough evidence to reject H_0.

A test result is significant when the P-value is small enough. Usually any value less than 0.05 (with a confidence interval of 95%). The outcome for the decision can be as per the Table A1.1 below.

Table A1.1: Hypothesis Testing

P – Value	Test Result	Action
< 0.05	Significant	Reject Ho in favour of H₁
> 0.05	Non-significant	Do not reject Ho

Testing can be done at any level of significance. Most researchers use a Confidence Interval of 95% which translates to 5% of significance.

If the hypothesis testing is needed for a critical test, such as life threatening than it is appropriate to use 1%. It is commonly used for safety related or pharmaceutical products.

APPENDIX TWO

MINITAB

MiniTab Computation.

Load the Minitab programme. Worksheet 1 will appear on screen.

Input lead contamination data in column *C1* - (*200, 180, 870, 150, 100, 120, 80, 50*). Click on Stat, then basic statistics, then 1 sample t-test.

Select '*Samples in Column*'. *C1 mg/kg* will appear. Click on it. Input in Test mean: 600

Select *Options*, than input in *C1: 95%*, than input in *Alternative: less than*, then click *OK*. Select Graphs, click on Individual Value Plot, than OK.

Click OK on '1 Sample t (Test and Confidence Interval).

Results will appear as

Table A2.1: Sample of Statistical Analysis by MiniTab

One-sample T-test: Test for mu = 30 vs < 30	Chromium in Groundwater (µg/L)						
Variable	N	Mean	St. Dev	SE Mean	95% Upper Bound	T	P
Chromium (µg/L)	13	30.69	16.45	4.56	38.82	0.15	0.56

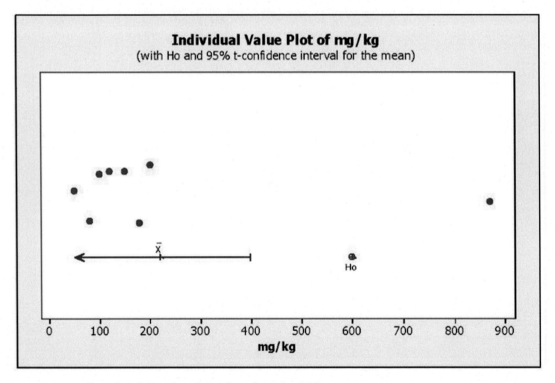

Figure A2.1: Sample of Graphical Analysis by MiniTab

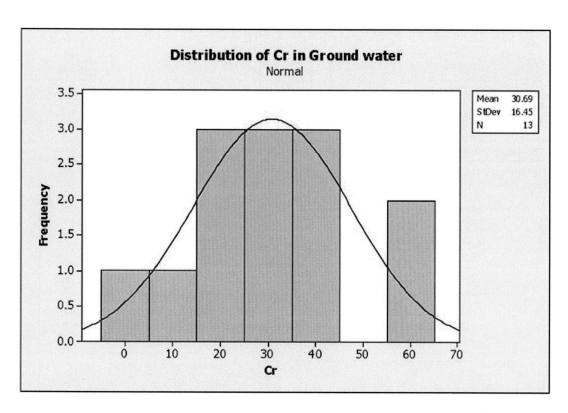

Figure A2.2: Distribution of Chromium in Groundwater

The mean value of the distribution is 30.7 and the standard deviation is 16.4 on the graph of distribution of chromium in groundwater as can be seen in Figure A2.2. The frequency of occurrence is normally distributed. There are two occurrence of 60 ppm.

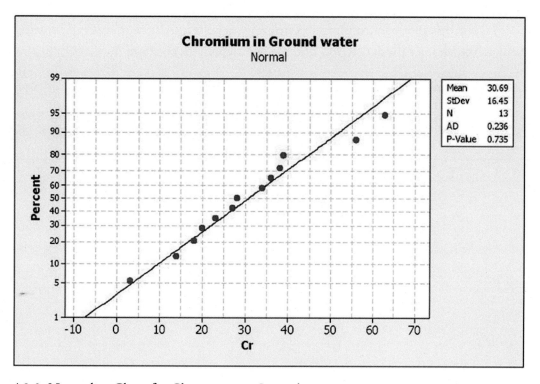

Figure A2.3: Normality Chart for Chromium in Groundwater

From Figure A2.3 the normality chart of chromium in groundwater, all the points are located around the straight line. The P-value is 0.74, meaning that the data is normal and the distribution is normally distributed.

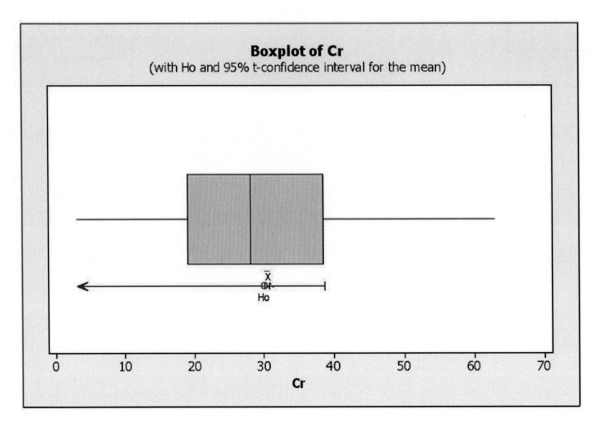

Figure 2.4 Box-plot of Chromium in Groundwater

Figure A2.4 is the Box-plot of chromium in Groundwater chart. All the points fall between 20 and 40 in the plot. The mean of the data is 30.7 and the null hypothesis equals to 30.

Table A2.2: One Sample T-test for Chromium in Groundwater.

One-sample T-test: Test for mu = 30 vs < 30	Chromium in Groundwater (µg/L)						
Variable	N	Mean	St. Dev	SE Mean	95% Upper Bound	T	P
Chromium (µg/L)	13	30.69	16.45	4.56	38.82	0.15	0.56

For hypothesis testing, the data was tested to a null hypothesis as equal to 30 at an alpha risk level of 5%. The P-value from the analysis was found to be 0.56 as can be seen from Table A2.2. Based on the P-value the null hypothesis is accepted and the alternative hypothesis is rejected. This shows that the mean value from the investigation is equal to the null hypothesis of 30.

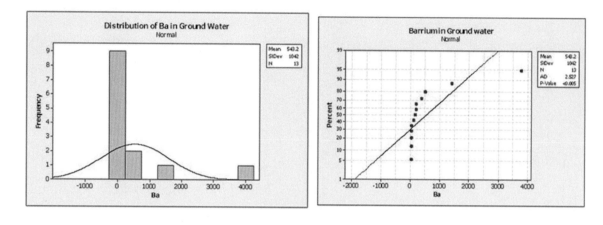

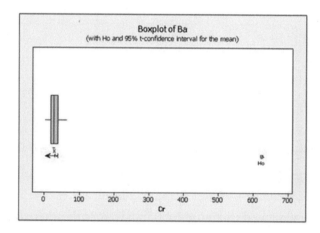

Figure A2.5: Graphical Analysis of Barium in Groundwater

For barium in groundwater the analysis is shown in Figure A2.5. There are two elevated values which are high, resulting in the standard deviation to be high. The normality chart is not normally distributed as the values do not fall on the straight line, indicating that the distribution is not normal. The P-value is <0.05, the null hypothesis is rejected in favour of the alternative hypothesis.

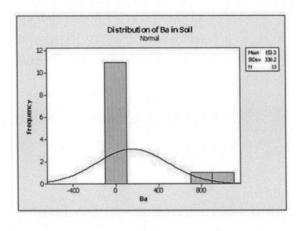
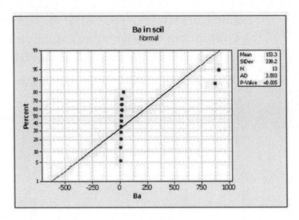
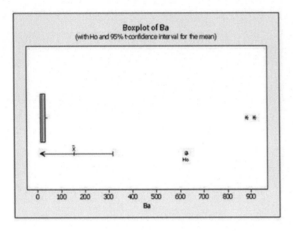
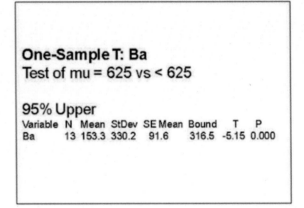

Figure A2.6: Graphical Analysis of Barium in Soil

For barium in soil, the mean was 153 ppm and the standard deviation was 330, indicating that there are two reading with elevated readings, as can be seen the box-plot graph in Figure A2.6. The null hypothesis is rejected as the P-value is <0.05, as evidenced in Figure A2.6 in favour of the alternative hypothesis.

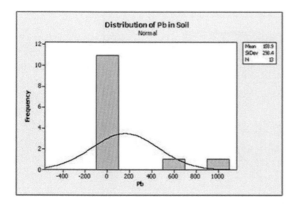
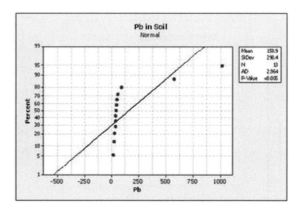
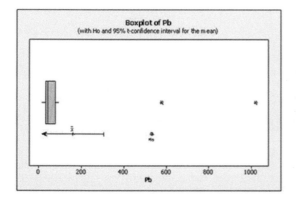

Figure A2.7: Graphical Analysis of Lead in Soil

The mean for lead in soil was 159 ppm as shown in the table in Figure A2.7 and the standard deviation is 298 ppm. The standard deviation is high due to two elevated readings. The P-value is <0.05, so the null hypothesis is rejected in favour of the alternative hypothesis.

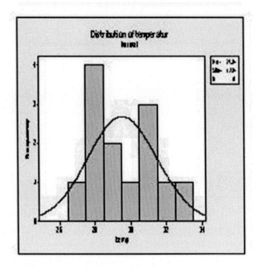

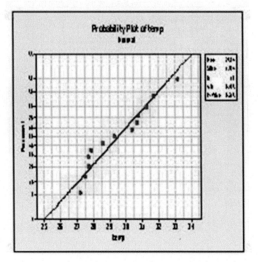

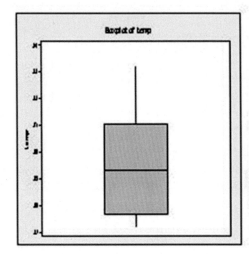

Q1 = 27.7
Median = 29.3
Q3 = 31.05
IQRange = 3.35
Whiskers to: 27.2, 33.2
N = 13

Figure A2.8: Graphical Analysis of Temperature in Groundwater

The temperature of groundwater is normally distributed as shown in the normality chart in Figure A2.8. The mean of groundwater temperature is 29.5 °C and the P-value is >0.05, for this case we cannot reject the null hypothesis.

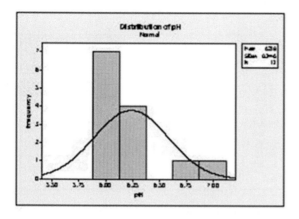
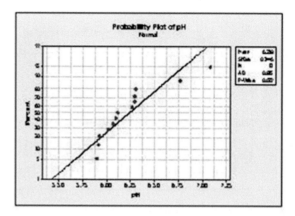
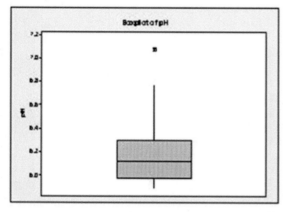

Q1 = 5.975
Median = 6.12
Q3 = 6.305
IQRange = 0.33
Whiskers to: 5.9, 6.77
N = 13

Figure A2.9: Graphical Analysis of pH in Groundwater

The pH of groundwater ranged between 5.90 and 6.77, there was one value at 7.08. The site is located close to the shore line, as such the groundwater could have been influenced by the sea water making it slightly acidic.

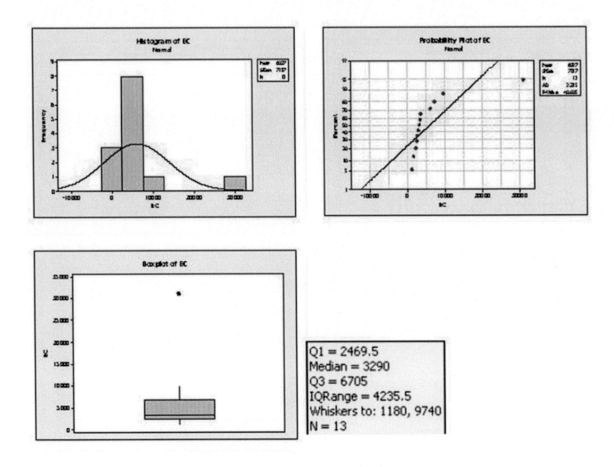

Figure A2.10: Graphical Analysis of Electrical Conductivity (EC) in Groundwater

The electrical conductivity of groundwater ranged between 1,180 and 9,740 µS/cm and there was one well which had 31,200 µS/cm. The median of the EC value is 3,290 as shown in Figure A2.10 box-plot.

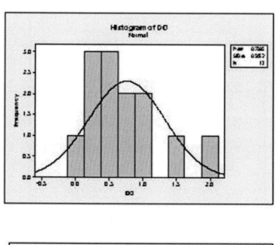
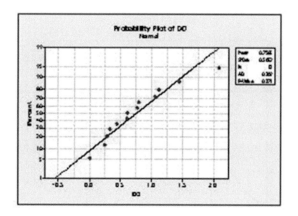
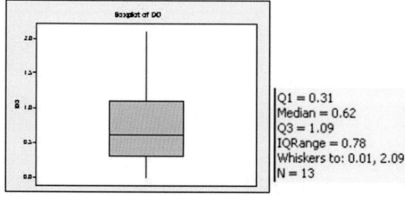

Q1 = 0.31
Median = 0.62
Q3 = 1.09
IQRange = 0.78
Whiskers to: 0.01, 2.09
N = 13

Figure A2.11: Graphical Analysis of Dissolved Oxygen (DO) in Groundwater

Dissolved oxygen in groundwater is normally distributed. The values are close to the normality chart as can be seen in Figure A2.11. Also, if the null hypothesis is tested for its mean to equal 0.76, it will have a P-value >0.05, meaning that if any further readings are evaluated they will fall within the same range.

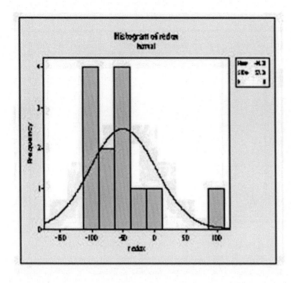
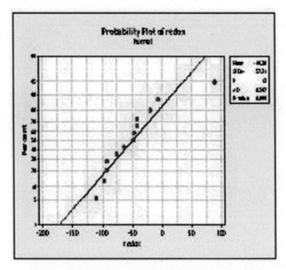
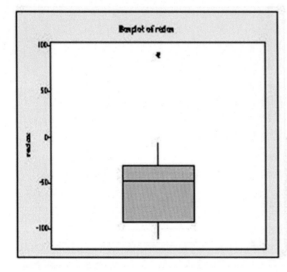

Q1 = -92
Median = -48
Q3 = -30
IQRange = 62
Whiskers to: -110, -6
N = 13

Figure A2.12: Graphical Analysis of Radox in Groundwater

The readings for radox are normally distributed as evidenced from the normality graph in Figure A2.12. The box-plot has only one reading that is above the range and it could be an outliner. The P-value is also >0.05, indicating the null hypothesis test will not be rejected when tested to equal to the mean.

APPENDIX THREE

MEASURING EQUIPMENT

Underground Cable Detector

Buried underground cables and piping are a challenge to electricians, plumbers and excavating jobbers. The excavating job also poses as a hazard to the crew. Improper digging can cause personal injury, property damage, costly repairs and penalties imposed by the local authorities. The equipment locates underground energised and non-energised cables, wires and pipes. The equipment provides an audible signal for detection purposes and visual display on its LCD screen. As such, screening for underground cables and piping becomes simpler and less costly. Countries like the US the technician operating such equipment have to be certified and licensed.

One such equipment is the Amprobe AT3000 PRO. The complete set cost about US$3500 (three thousand five hundred dollars). It is important to note that when selecting such an equipment to what depth it is sensitive to.

Photo Ionisation Detector (PID)

A PID measures VOCs and other toxic gases in low concentrations. This is a sensitive broad spectrum monitor. The PID uses an ultraviolet light source to break down chemicals to positive and negative ions that can easily be counted with a detector. The ionization process occurs when a molecule absorbs the high energy ultraviolet light, which will excite the molecule and results in the temporary loss of a negatively charged electron and the formation of positively charged ion. The gas becomes electrically charged. These charged particles produce a current that is then amplified and displayed on the instrument. The electrons will recombine after the electrodes are removed to their original molecular form. The equipment is non-destructive during the testing period and as such further testing can be conducted immediately.

All element and chemicals can be ionised, but they defer in their energy levels and the energy they require during the process. The energy required to displace an electron and ionise a compound is called its Ionization Potential (IP) and it is measured in electron volt (eV). The light energy emitted by the ultraviolet lamp is also measured in eV. If the IP sample gas is less than the eV output of the lamp, than the sample gas will be ionized. The PID uses the lamp to break down gases and vapours. Each gas has its own standard PID lamp. As such it is appropriate to select the lamps to determine the gases under investigation. The readings from the monitor can be summarised as follows; If the eV of a gas or vapour is less than the eV of the PID lamp, than the PID can see the gas or vapour. If the eV of a gas or vapour is greater than the eV of the PID lamp, than the PID cannot see the gas or vapour. The PID are calibrated using isobutylene gas and they will indicate how much of a gas or vapour is present. Using the placard or manifest the chemical need to be identified. The sensitivity of the instrument is adjusted to the chemical and the reading from the instrument is recorded. Some instruments use correction factors which allow to measure more than one chemical with one calibration. The PID instrument used was the ToxiRAE Plus PID. This is a light weight and compact VOC gas monitor. The cost of the equipment is about US$4500 (four thousand five hundred dollars).

APPENDIX FOUR

MONITORING WELLS

Piezometer

Principally, it is used to measure the pressure of a fluid at a specific location in a column. This determines the water level below ground and or a datum point. Piezometers have a short screen and filter zone, this allows it to represent the hydraulic head at a point in the aquifer. Piezometer are sometimes or most often referred to as monitoring wells. Monitoring well are often smaller diameter wells used to monitor the hydraulic head or sample the groundwater for chemical constituents. Piezometers are monitoring wells completed over a very short section of aquifer. Monitoring wells can also be completed at multiple levels, allowing discrete samples or measurements to be made at different vertical elevations at the same map location.

As water percolates into the aquifer and flows downhill, the lower layers come under pressure from the water that flows down. The greater the vertical distance between the recharge area and the bottom area, the higher the pressure will be. If small wells are placed along the aquifer the water pressure will cause water to rise to a place of equilibrium. The depth of the water level from the ground level and a reference datum will indicate the flow of water. A plane drawn through the points of equilibrium is called a piezometric surface. In an unconfined aquifer the piezometric surface is the water table. As such, the water surface slopes from areas of recharge to areas of discharge. The pressure difference represented by the slopes cause the flow of water within the aquifer. At any point the slope is a reflection of the rate of flow of water and the resistance to movement of water through the ground.

If the surface lies above the ground surface, water will flow from the well naturally. When the surface is below the ground surface, water will have to be pumped from the well.

Flow of water is based on the principle of Darcy's Law (Mackenzie and Cornwell, 1991) which states that the amount of groundwater discharging through a given portion of an aquifer is proportional to the cross section area to flow and the hydraulic head gradient.

Plate A4.2: Man Cleaning a Well

As indicated by Walter (1982), in countries like Cameroon or some rural areas of Malaysia drinking water are obtained from wells. Shallow wells can be constructed at very low cost but has a greater risk of contamination compared to deeper wells. This is due to impurities from the surface can easily reach shallow sources. As such, water from the artesian wells, rises to a greater level than the land surface when extracted from a deep source. Well water for personal consumption is often filtered with water processors. The processors remove small particles from the water, after which the water is boiled to kill the microorganisms. This situation is still common in rural areas of under developed countries. The picture in Figure A4.2 above shows a man is clearing the well from contamination. The water is drained to remove debris and small particles so that fresh water from the ground could be received, thus allowing the users to obtain fresh water for consumption.

Contamination by human activities is a common problem with ground water in urban areas. They are mostly due to gasoline spills and underground storage tank leakages and the ingredients are commonly benzene, toluene, ethylbenzene and xylene. Also, many industrial solvents are common underground contaminants. They enter groundwater through leaks, accidental spills or intentional dumping.

Clean-up of contaminated groundwater tends to be very costly and effective remediation of groundwater is generally very difficult.

APPENDIX FIVE

SURFER SIX

Surfer software is a tool that is used for contouring, gridding and surface mapping. The software runs on Microsoft Windows. The Surfer programme converts the input data into contours, vector flows and gridding. There are other features in co-operated as well.

Vector Maps

The programme instantly creates vector maps in Surfer to show direction and magnitude of data at points on a map. It creates vector maps from information in one grid or two separate grids. The two components of the vector map, direction and magnitude, are automatically generated from a single grid by computing the gradient of the represented surface. At any given grid node, the direction of the arrow points in the direction of the steepest descent. The magnitude of the arrow changes depending on the steepness of the descent. Two-grid vector maps use two separate grid files to determine the vector direction and magnitude. The grids can contain Cartesian or polar data. With Cartesian data, one grid consists of X component data and the other grid consists of Y component data. With polar data, one grid consists of angle information and the other grid contains length information. Vector maps can be over lay on contours to enhance the visualisation of the status.

Contour Maps

Surfer software's contour maps give a full control over all map parameters. Surfer's intelligent defaults automatically create a contour map, or on a double-click on the mouse a map is easily customized to its map features. It displays contour maps over any contour range and contour interval, or the user need to specify only the contour levels to be displayed on the map. The Surfer software adds colour fill between contours to produce dazzling displays of maps, or produce grey scale fills for black and white printouts.

Gridding

The gridding methods in Surfer allows to produce accurate contour, surface, wire frame, vector, image, and shaded relief maps from XYZ data. The data can be randomly dispersed over the map area, and Surfer's gridding will interpolate the data onto a grid. There are a multitude of gridding methods to choose from, allowing the ability to produce exactly the map of choice. With each gridding method you have complete control over the gridding parameters. If your data are already collected in a regular rectangular array, you can create a map directly from your data. Computer generated contour maps have never been more accurate.

The cost of a single stand-alone license is US$550.00 and the cost of an academic license is US$495.00 for the Surfer software. It can be purchased from Rockware Inc.

APPENDIX SIX

LABORATORY RESULTS

An accredited laboratory was used to analyse the investigative samples collected from the site. This laboratory was selected due to its accreditation, service and cost competiveness.

A summary of results has been extracts and are shown in the following tables in the subsequent pages.

Table A6.1: Submission of samples for testing – Certificate of Analysis.

SAMPLE I.D	MATRIX	TIME (hrs)	REMARKS
BH4/1.7	Soil	1050	-
BH5/1.5	Soil	1113	-
BH6/1.5	Soil	1216	-
BH7/1.5	Soil	1427	-
BH8/1.5	Soil	1518	-
BH9/1.5	Soil	1648	-
BH10/1.5	Soil	0814	-
BH11/1.5	Soil	0903	-
BH12/1.5	Soil	1143	-
BH13/1.7	Soil	1440	-
BH4/GW	Water	0915	-
BH5/GW	Water	0947	-
BH6/GW	Water	1032	-
BH7/GW	Water	1109	-
BH8/GW	Water	1130	-
BH9/GW	Water	1156	-
BH10/GW	Water	1220	-
BH11/GW	Water	1243	-
BH12/GW	Water	1410	-
BH13/GW	Water	1440	-
Dup-I	Soil	-	-
Dup-II	Water	-	-

Table A6.2: Test Results of Metals in Soil

Method Reference	Analysis Description	Lab I.D		8232 BH4/1.7	8233 BH5/1.5	8234 BH6/1.5	8235 BH7/1.5	8236 BH8/1.5	8237 BH9/1.5
		Sample I.D							
		Units	LOR						
		Date of Digestion		4/4/05					
		Date of Analysis		5/4/05					
APHA 2540 G	Moisture @ 103 °C	%	0.1	33.2	26.3	19.1	23.0	17.6	20.0
USEPA 3050B, 6010B	Arsenic	mg/kg	5	<5	<5	<5	<5	11	<5
USEPA 3050B, 6010B	Cadmium	mg/kg	1	<1	<1	<1	<1	<1	<1
USEPA 3050B, 6010B	Chromium	mg/kg	1	11	2	2	3	2	3
USEPA 3050B, 6010B	Lead	mg/kg	1	25	94	45	36	1020	576
USEPA 3050B, 6010B	Selenium	mg/kg	10	<10	<10	<10	<10	<10	<10
USEPA 3050B, 6010B	Silver	mg/kg	1	<1	<1	<1	<1	<1	<1
USEPA 3050B, 6010B	Barium	mg/kg	10	12	40	15	16	914	880
USEPA 7471 A	Mercury	mg/kg	0.5	<0.5	<0.5	<0.5	<0.5	<0.5	<0.5

Method Reference	Analysis Description	Lab I.D		8238 BH10/1.5	8239 BH11/1.5	8240 BH12/1.5	8241 BH13/1.7	8252 Dup-I
		Sample I.D						
		Units	LOR					
		Date of Digestion						
		Date of Analysis						
APHA 2540 G	Moisture @ 103 °C	%	0.1	20.0	20.9	19.6	21.5	22.6
USEPA 3050B, 6010B	Arsenic	mg/kg	5	<5	<5	<5	<5	<5
USEPA 3050B, 6010B	Cadmium	mg/kg	1	<1	<1	<1	<1	<1
USEPA 3050B, 6010B	Chromium	mg/kg	1	1	2	1	3	2
USEPA 3050B, 6010B	Lead	mg/kg	1	39	29	60	51	64
USEPA 3050B, 6010B	Selenium	mg/kg	10	<10	<10	<10	<10	<10
USEPA 3050B, 6010B	Silver	mg/kg	1	<1	<1	<1	<1	<1
USEPA 3050B, 6010B	Barium	mg/kg	10	22	19	22	20	28
USEPA 7471 A	Mercury	mg/kg	0.5	<0.5	<0.5	<0.5	<0.5	<0.5

LOR: Level of Reporting

Table A6.3: Test results of VOC in Soil

Method Reference	Analysis Description	Units	LOR	8232 BH4/1.7	8233 BH5/1.5	8234 BH6/1.5	8235 BH7/1.5	8236 BH8/1.5	8237 BH9/1.5
	Lab I.D								
	Sample I.D								
	Date of Analysis						5/4/05		
USEPA 5030B, 8260B	Volatile Organic Carbon (VOC)								
USEPA 5030B, 8260B	Monocylic Aromatics								
USEPA 5030B, 8260B	Benzene	mg/kg	0.5	<0.5	<0.5	<0.5	<0.5	<0.5	<0.5
USEPA 5030B, 8260B	Toluene	mg/kg	0.5	<0.5	<0.5	<0.5	<0.5	<0.5	<0.5
USEPA 5030B, 8260B	Ethylbenzene	mg/kg	0.5	<0.5	<0.5	<0.5	<0.5	<0.5	<0.5
USEPA 5030B, 8260B	m & p-Xylene	mg/kg	1	<1	<1	<1	<1	<1	<1
USEPA 5030B, 8260B	Styrene	mg/kg	0.5	<0.5	<0.5	<0.5	<0.5	<0.5	<0.5
USEPA 5030B, 8260B	o-Xylene	mg/kg	0.5	<0.5	<0.5	<0.5	<0.5	<0.5	<0.5
USEPA 5030B, 8260B	Isopropylbenzene	mg/kg	0.5	<0.5	<0.5	<0.5	<0.5	<0.5	<0.5
USEPA 5030B, 8260B	n-Propylbenzene	mg/kg	0.5	<0.5	<0.5	<0.5	<0.5	<0.5	<0.5
USEPA 5030B, 8260B	1,3,5-Trimethylbenzene	mg/kg	0.5	<0.5	<0.5	<0.5	<0.5	<0.5	<0.5
USEPA 5030B, 8260B	sec-Butylbenzene	mg/kg	0.5	<0.5	<0.5	<0.5	<0.5	<0.5	<0.5
USEPA 5030B, 8260B	1,2,4-Trimethylbenzene	mg/kg	0.5	<0.5	<0.5	<0.5	<0.5	<0.5	<0.5
USEPA 5030B, 8260B	tert-Butylbenzene	mg/kg	0.5	<0.5	<0.5	<0.5	<0.5	<0.5	<0.5
USEPA 5030B, 8260B	p-Isopropyltoluene	mg/kg	0.5	<0.5	<0.5	<0.5	<0.5	<0.5	<0.5
USEPA 5030B, 8260B	n-Butylbenzene	mg/kg	0.5	<0.5	<0.5	<0.5	<0.5	<0.5	<0.5
USEPA 5030B, 8260B	Oxygenated Compounds								
USEPA 5030B, 8260B	2-Butanone (MEK)	mg/kg	5	<5	<5	<5	<5	<5	<5
USEPA 5030B, 8260B	4-Methyl-2-pentanone (MIBK)	mg/kg	5	<5	<5	<5	<5	<5	<5
USEPA 5030B, 8260B	2-Hexanone (MBK)	mg/kg	5	<5	<5	<5	<5	<5	<5
USEPA 5030B, 8260B	Fumigants								
USEPA 5030B, 8260B	2,2-Dichloropropane	mg/kg	0.5	<0.5	<0.5	<0.5	<0.5	<0.5	<0.5
USEPA 5030B, 8260B	1,2-Dichloropropane	mg/kg	0.5	<0.5	<0.5	<0.5	<0.5	<0.5	<0.5
USEPA 5030B, 8260B	cis-1,3-Dichloropropylene	mg/kg	0.5	<0.5	<0.5	<0.5	<0.5	<0.5	<0.5
USEPA 5030B, 8260B	trans-1,3-Dichloropropylene	mg/kg	0.5	<0.5	<0.5	<0.5	<0.5	<0.5	<0.5
USEPA 5030B, 8260B	1,2-Dibromoethane	mg/kg	0.5	<0.5	<0.5	<0.5	<0.5	<0.5	<0.5
USEPA 5030B, 8260B	Halogenated Aliphatics								
USEPA 5030B, 8260B	Dichlorodifluoromethane	mg/kg	5	<5	<5	<5	<5	<5	<5
USEPA 5030B, 8260B	Chloromethane	mg/kg	5	<5	<5	<5	<5	<5	<5
USEPA 5030B, 8260B	Vinyl chloride	mg/kg	5	<5	<5	<5	<5	<5	<5
USEPA 5030B, 8260B	Bromomethane	mg/kg	5	<5	<5	<5	<5	<5	<5
USEPA 5030B, 8260B	Chloroethane	mg/kg	5	<5	<5	<5	<5	<5	<5
USEPA 5030B, 8260B	Trichlorofluoromethane	mg/kg	5	<5	<5	<5	<5	<5	<5
USEPA 5030B, 8260B	1,1-Dichloroethylene	mg/kg	0.2	<0.2	<0.2	<0.2	<0.2	<0.2	<0.2
USEPA 5030B, 8260B	trans-1,2-Dichloroethylene	mg/kg	0.5	<0.5	<0.5	<0.5	<0.5	<0.5	<0.5
USEPA 5030B, 8260B	1,1-Dichloroethane	mg/kg	0.5	<0.5	<0.5	<0.5	<0.5	<0.5	<0.5

LOR: Level of Reporting

Table A6.3: Continued

Method Reference	Analysis Description	Units	LOR	BH4/G W	BH5/G W	BH6/G W	BH7/G W	BH8/G W	BH9/G W
	Lab I.D			8242	8243	8244	8245	8246	8247
	Date of Analysis			4/4/05					
USEPA 5030B, 8260B	Volatile Organic Carbon (VOC)								
USEPA 5030B, 8260B	Monocylic Aromatics								
USEPA 5030B, 8260B	Benzene	ug/l	5	<5	<5	<5	<5	<5	<5
USEPA 5030B, 8260B	Toluene	ug/l	5	<5	<5	<5	<5	<5	<5
USEPA 5030B, 8260B	Ethylbenzene	ug/l	5	<5	<5	<5	<5	<5	<5
USEPA 5030B, 8260B	m & p-Xylene	ug/l	10	<10	<10	<10	<10	<10	<10
USEPA 5030B, 8260B	Styrene	ug/l	5	<5	<5	<5	<5	<5	<5
USEPA 5030B, 8260B	o-Xylene	ug/l	5	<5	<5	<5	<5	<5	<5
USEPA 5030B, 8260B	Isopropylbenzene	ug/l	5	<5	<5	<5	<5	<5	<5
USEPA 5030B, 8260B	n-Propylbenzene	ug/l	5	<5	<5	<5	<5	<5	<5
USEPA 5030B, 8260B	1,3,5-Trimethylbenzene	ug/l	5	<5	<5	<5	<5	<5	<5
USEPA 5030B, 8260B	sec-Butylbenzene	ug/l	5	<5	<5	<5	<5	<5	<5
USEPA 5030B, 8260B	1,2,4-Trimethylbenzene	ug/l	5	<5	<5	<5	<5	<5	<5
USEPA 5030B, 8260B	tert-Butylbenzene	ug/l	5	<5	<5	<5	<5	<5	<5
USEPA 5030B, 8260B	p-Isopropyltoluene	ug/l	5	<5	<5	<5	<5	<5	<5
USEPA 5030B, 8260B	n-Butylbenzene	ug/l	5	<5	<5	<5	<5	<5	<5
USEPA 5030B, 8260B	Oxygenated Compounds								
USEPA 5030B, 8260B	2-Butanone (MEK)	ug/l	50	<50	<50	<50	<50	<50	<50
USEPA 5030B, 8260B	4-Methyl-2-pentanone (MIBK)	ug/l	50	<50	<50	<50	<50	<50	<50
USEPA 5030B, 8260B	2-Hexanone (MBK)	ug/l	50	<50	<50	<50	<50	<50	<50
USEPA 5030B, 8260B	Fumigants								
USEPA 5030B, 8260B	2,2-Dichloropropane	ug/l	5	<5	<5	<5	<5	<5	<5
USEPA 5030B, 8260B	1,2-Dichloropropane	ug/l	5	<5	<5	<5	<5	<5	<5
USEPA 5030B, 8260B	cis-1,3-Dichloropropylene	ug/l	5	<5	<5	<5	<5	<5	<5
USEPA 5030B, 8260B	trans-1,3-Dichloropropylene	ug/l	5	<5	<5	<5	<5	<5	<5
USEPA 5030B, 8260B	1,2-Dibromoethane	ug/l	5	<5	<5	<5	<5	<5	<5
USEPA 5030B, 8260B	Halogenated Aliphatics								
USEPA 5030B, 8260B	Dichlorodifluoromethane	ug/l	50	<50	<50	<50	<50	<50	<50
USEPA 5030B, 8260B	Chloromethane	ug/l	50	<50	<50	<50	<50	<50	<50
USEPA 5030B, 8260B	Vinyl chloride	ug/l	50	<50	<50	<50	<50	<50	<50
USEPA 5030B, 8260B	Bromomethane	ug/l	50	<50	<50	<50	<50	<50	<50
USEPA 5030B, 8260B	Chloroethane	ug/l	50	<50	<50	<50	<50	<50	<50
USEPA 5030B, 8260B	Trichlorofluoromethane	ug/l	50	<50	<50	<50	<50	<50	<50
USEPA 5030B, 8260B	1,1-Dichloroethylene	ug/l	5	<5	<5	<5	<5	<5	<5
USEPA 5030B, 8260B	trans-1,2-Dichloroethylene	ug/l	5	<5	<5	<5	<5	<5	<5
USEPA 5030B, 8260B	1,1-Dichloroethane	ug/l	5	<5	<5	<5	<5	<5	<5

LOR: Level of Reporting

Table A6.4: Test Results of Metals in Groundwater

Method Reference	Analysis Description	Lab I.D		8242	8243	8244	8245	8246	8247
		Sample I.D		BH4/G	BH5/G	BH6/G	BH7/G	BH8/G	BH9/G
		Units	LOR	W	W	W	W	W	W
USEPA 6010B	Arsenic	ug/l	50	<50	<50	<50	<50	<50	<50
USEPA 6010B	Cadmium	ug/l	1	<1	<1	<1	<1	<1	<1
USEPA 6010B	Chromium	ug/l	1	28	34	63	38	18	27
USEPA 6010B	Lead	ug/l	10	<10	<10	<10	<10	<10	<10
USEPA 6010B	Selenium	ug/l	100	<100	<100	<100	<100	<100	<100
USEPA 6010B	Silver	ug/l	10	<10	<10	<10	<10	<10	<10
USEPA 6010B	Barium	ug/l	50	<50	<50	<50	<50	1420	510
USEPA 7196A	Chromium Hexavalent	ug/l	50	<50	<50	<50	<50	<50	<50
USEPA 7470 A	Mercury	ug/l	0.2	<0.2	<0.2	<0.2	<0.2	<0.2	<0.2

Method Reference	Analysis Description	Lab I.D		8248	8249	8250	8251	8253	
		Sample I.D		BH10/G	BH11/G	BH12/G	BH13/G	Dup-II	
		Units	LOR	W	W	W	W		
USEPA 6010B	Arsenic	ug/l	50	<50	<50	<50	<50	<50	
USEPA 6010B	Cadmium	ug/l	1	<1	<1	<1	<1	<1	
USEPA 6010B	Chromium	ug/l	1	39	56	23	36	35	
USEPA 6010B	Lead	ug/l	10	<10	<10	<10	<10	<10	
USEPA 6010B	Selenium	ug/l	100	<100	<100	<100	<100	<100	
USEPA 6010B	Silver	ug/l	10	<10	<10	<10	<10	<10	
USEPA 6010B	Barium	ug/l	50	206	199	173	392	<50	
USEPA 7196A	Chromium Hexavalent	ug/l	50	<50	<50	<50	<50	<50	
USEPA 7470 A	Mercury	ug/l	0.2	<0.2	<0.2	<0.2	<0.2	<0.2	

LOR: Level of Reporting

Table A6.5: Test Results of VOC in Groundwater

Method Reference	Analysis Description	Units	LOR	8242 BH4/G W	8243 BH5/G W	8244 BH6/G W	8245 BH7/G W	8246 BH8/G W	8247 BH9/G W
	Lab I.D / Sample I.D								
Method Reference	Analysis Description	Date of Analysis				4/4/05			
USEPA 5030B, 8260B	Volatile Organic Carbon (VOC)								
USEPA 5030B, 8260B	Monocylic Aromatics								
USEPA 5030B, 8260B	Benzene	ug/l	5	<5	<5	<5	<5	<5	<5
USEPA 5030B, 8260B	Toluene	ug/l	5	<5	<5	<5	<5	<5	<5
USEPA 5030B, 8260B	Ethylbenzene	ug/l	5	<5	<5	<5	<5	<5	<5
USEPA 5030B, 8260B	m & p-Xylene	ug/l	10	<10	<10	<10	<10	<10	<10
USEPA 5030B, 8260B	Styrene	ug/l	5	<5	<5	<5	<5	<5	<5
USEPA 5030B, 8260B	o-Xylene	ug/l	5	<5	<5	<5	<5	<5	<5
USEPA 5030B, 8260B	Isopropylbenzene	ug/l	5	<5	<5	<5	<5	<5	<5
USEPA 5030B, 8260B	n-Propylbenzene	ug/l	5	<5	<5	<5	<5	<5	<5
USEPA 5030B, 8260B	1,3,5-Trimethylbenzene	ug/l	5	<5	<5	<5	<5	<5	<5
USEPA 5030B, 8260B	sec-Butylbenzene	ug/l	5	<5	<5	<5	<5	<5	<5
USEPA 5030B, 8260B	1,2,4-Trimethylbenzene	ug/l	5	<5	<5	<5	<5	<5	<5
USEPA 5030B, 8260B	tert-Butylbenzene	ug/l	5	<5	<5	<5	<5	<5	<5
USEPA 5030B, 8260B	p-Isopropyltoluene	ug/l	5	<5	<5	<5	<5	<5	<5
USEPA 5030B, 8260B	n-Butylbenzene	ug/l	5	<5	<5	<5	<5	<5	<5
USEPA 5030B, 8260B	Oxygenated Compounds								
USEPA 5030B, 8260B	2-Butanone (MEK)	ug/l	50	<50	<50	<50	<50	<50	<50
USEPA 5030B, 8260B	4-Methyl-2-pentanone (MIBK)	ug/l	50	<50	<50	<50	<50	<50	<50
USEPA 5030B, 8260B	2-Hexanone (MBK)	ug/l	50	<50	<50	<50	<50	<50	<50
USEPA 5030B, 8260B	Fumigants								
USEPA 5030B, 8260B	2,2-Dichloropropane	ug/l	5	<5	<5	<5	<5	<5	<5
USEPA 5030B, 8260B	1,2-Dichloropropane	ug/l	5	<5	<5	<5	<5	<5	<5
USEPA 5030B, 8260B	cis-1,3-Dichloropropylene	ug/l	5	<5	<5	<5	<5	<5	<5
USEPA 5030B, 8260B	trans-1,3-Dichloropropylene	ug/l	5	<5	<5	<5	<5	<5	<5
USEPA 5030B, 8260B	1,2-Dibromoethane	ug/l	5	<5	<5	<5	<5	<5	<5
USEPA 5030B, 8260B	Halogenated Aliphatics								
USEPA 5030B, 8260B	Dichlorodifluoromethane	ug/l	50	<50	<50	<50	<50	<50	<50
USEPA 5030B, 8260B	Chloromethane	ug/l	50	<50	<50	<50	<50	<50	<50
USEPA 5030B, 8260B	Vinyl chloride	ug/l	50	<50	<50	<50	<50	<50	<50
USEPA 5030B, 8260B	Bromomethane	ug/l	50	<50	<50	<50	<50	<50	<50
USEPA 5030B, 8260B	Chloroethane	ug/l	50	<50	<50	<50	<50	<50	<50
USEPA 5030B, 8260B	Trichlorofluoromethane	ug/l	50	<50	<50	<50	<50	<50	<50
USEPA 5030B, 8260B	1,1-Dichloroethylene	ug/l	5	<5	<5	<5	<5	<5	<5
USEPA 5030B, 8260B	trans-1,2-Dichloroethylene	ug/l	5	<5	<5	<5	<5	<5	<5
USEPA 5030B, 8260B	1,1-Dichloroethane	ug/l	5	<5	<5	<5	<5	<5	<5

LOR: Level of Reporting

Table A6.5: Continued

			Lab I.D		8242	8243	8244	8245	8246	8247
			Sample I.D		BH4/G	BH5/G	BH6/G	BH7/G	BH8/G	BH9/G
		Units	LOR		W	W	W	W	W	W
USEPA 5030B, 8260B	cis-1,2-Dichloroethylene	ug/l	5		<5	<5	<5	<5	<5	<5
USEPA 5030B, 8260B	1,1,1-Trichloroethane	ug/l	5		<5	<5	<5	<5	<5	<5
USEPA 5030B, 8260B	1,1-Dichloropropylene	ug/l	5		<5	<5	<5	<5	<5	<5
USEPA 5030B, 8260B	Carbon tetrachloride	ug/l	5		<5	<5	<5	<5	<5	<5
USEPA 5030B, 8260B	1,2-Dichloroethane	ug/l	5		<5	<5	<5	<5	<5	<5
USEPA 5030B, 8260B	Trichloroethylene	ug/l	5		<5	<5	<5	<5	<5	<5
USEPA 5030B, 8260B	Dibromomethane	ug/l	5		<5	<5	<5	<5	<5	<5
USEPA 5030B, 8260B	1,1,2-Trichloroethane	ug/l	5		<5	<5	<5	<5	<5	<5
USEPA 5030B, 8260B	1,3-Dichloropropane	ug/l	5		<5	<5	<5	<5	<5	<5
USEPA 5030B, 8260B	Tetrachloroethylene	ug/l	5		<5	<5	<5	<5	<5	<5
USEPA 5030B, 8260B	1,1,1,2-Tetrachloroethane	ug/l	5		<5	<5	<5	<5	<5	<5
USEPA 5030B, 8260B	1,1,2,2-Tetrachloroethane	ug/l	5		<5	<5	<5	<5	<5	<5
USEPA 5030B, 8260B	1,2,3-Trichloropropane	ug/l	5		<5	<5	<5	<5	<5	<5
USEPA 5030B, 8260B	1,2-Dibromo-3-chloropropane	ug/l	5		<5	<5	<5	<5	<5	<5
USEPA 5030B, 8260B	Hexachlorobutadiene	ug/l	5		<5	<5	<5	<5	<5	<5
USEPA 5030B, 8260B	Chlorobenzene	ug/l	5		<5	<5	<5	<5	<5	<5
USEPA 5030B, 8260B	Bromobenzene	ug/l	5		<5	<5	<5	<5	<5	<5
USEPA 5030B, 8260B	2-Chlorotoluene	ug/l	5		<5	<5	<5	<5	<5	<5
USEPA 5030B, 8260B	4-Chlorotoluene	ug/l	5		<5	<5	<5	<5	<5	<5
USEPA 5030B, 8260B	1,3-Dichlorobenzene	ug/l	5		<5	<5	<5	<5	<5	<5
USEPA 5030B, 8260B	1,4-Dichlorobenzene	ug/l	5		<5	<5	<5	<5	<5	<5
USEPA 5030B, 8260B	1,2-Dichlorobenzene	ug/l	5		<5	<5	<5	<5	<5	<5
USEPA 5030B, 8260B	1,2,4-Trichlorobenzene	ug/l	5		<5	<5	<5	<5	<5	<5
USEPA 5030B, 8260B	1,2,3-Trichlorobenzene	ug/l	5		<5	<5	<5	<5	<5	<5
USEPA 5030B, 8260B	**Trihalomethanes**									
USEPA 5030B, 8260B	Chloroform	ug/l	20		<20	<20	<20	<20	<20	<20
USEPA 5030B, 8260B	Bromodichloromethane	ug/l	5		<5	<5	<5	<5	<5	<5
USEPA 5030B, 8260B	Dibromochloromethane	ug/l	5		<5	<5	<5	<5	<5	<5
USEPA 5030B, 8260B	Bromoform	ug/l	5		<5	<5	<5	<5	<5	<5
USEPA 5030B, 8260B	**Surrogate**	Units	Limits							
USEPA 5030B, 8260B	1,2-Dichloroethane-d4	%	84-142		134	132	130	132	132	132
USEPA 5030B, 8260B	Toluene-d8	%	79-124		100	102	104	106	102	102
USEPA 5030B, 8260B	4-Bromofluorobenzene	%	71-123		80	80	80	80	78	82

LOR: Level of Reporting

103

Table A6.6: Test Results of sVOC in Groundwater

Method Reference	Analysis Description	Lab I.D		8242	8243	8244	8245	8246	8247
		Sample I.D		BH4/G	BH5/G	BH6/G	BH7/G	BH8/G	BH9/G
		Units	LOR	W	W	W	W	W	W
		Date of Extraction		5/4/2005					
		Date of Analysis		5/4/2005					
USEPA 3510C, 8270 C	Semivolatile Organic Carbon (SVOC)								
USEPA 3510C, 8270 C	Phenols								
USEPA 3510C, 8270 C	Phenol	ug/l	5	<5	<5	<5	<5	<5	<5
USEPA 3510C, 8270 C	2-chlorophenol	ug/l	5	<5	<5	<5	<5	<5	<5
USEPA 3510C, 8270 C	2-methylphenol	ug/l	5	<5	<5	<5	<5	<5	<5
USEPA 3510C, 8270 C	4-methylphenol	ug/l	5	<5	<5	<5	<5	<5	<5
USEPA 3510C, 8270 C	2-nitrophenol	ug/l	5	<5	<5	<5	<5	<5	<5
USEPA 3510C, 8270 C	2,4-dimethylphenol	ug/l	5	<5	<5	<5	<5	<5	<5
USEPA 3510C, 8270 C	2,4-dichlorophenol	ug/l	5	<5	<5	<5	<5	<5	<5
USEPA 3510C, 8270 C	2,6-dichlorophenol	ug/l	5	<5	<5	<5	<5	<5	<5
USEPA 3510C, 8270 C	4-chloro-3-methylphenol	ug/l	5	<5	<5	<5	<5	<5	<5
USEPA 3510C, 8270 C	2,4,6-trichlorophenol	ug/l	5	<5	<5	<5	<5	<5	<5
USEPA 3510C, 8270 C	2,4,5-trichlorophenol	ug/l	5	<5	<5	<5	<5	<5	<5
USEPA 3510C, 8270 C	Pentachlorophenol	ug/l	1	<1	<1	<1	<1	<1	<1
USEPA 3510C, 8270 C	Polynuclear Aromatics (PAH)								
USEPA 3510C, 8270 C	Naphthalene	ug/l	5	<5	<5	<5	<5	<5	<5
USEPA 3510C, 8270 C	Acenaphthylene	ug/l	5	<5	<5	<5	<5	<5	<5
USEPA 3510C, 8270 C	Acenaphthene	ug/l	5	<5	<5	<5	<5	<5	<5
USEPA 3510C, 8270 C	Fluorene	ug/l	5	<5	<5	<5	<5	<5	<5
USEPA 3510C, 8270 C	Phenanthrene	ug/l	5	<5	<5	<5	<5	<5	<5
USEPA 3510C, 8270 C	Anthracene	ug/l	5	<5	<5	<5	<5	<5	<5
USEPA 3510C, 8270 C	Fluoranthene	ug/l	5	<5	<5	<5	<5	<5	<5
USEPA 3510C, 8270 C	Pyrene	ug/l	5	<5	<5	<5	<5	<5	<5
USEPA 3510C, 8270 C	Benz(a)anthracene	ug/l	5	<5	<5	<5	<5	<5	<5
USEPA 3510C, 8270 C	Chrysene	ug/l	5	<5	<5	<5	<5	<5	<5
USEPA 3510C, 8270 C	Benzo(b) & (k)fluoranthene	ug/l	10	<10	<10	<10	<10	<10	<10
USEPA 3510C, 8270 C	Benzo(a)pyrene	ug/l	5	<5	<5	<5	<5	<5	<5
USEPA 3510C, 8270 C	Indeno (1,2,3-cd)pyrene	ug/l	5	<5	<5	<5	<5	<5	<5
USEPA 3510C, 8270 C	Dibenz(a,h)antharacene	ug/l	5	<5	<5	<5	<5	<5	<5
USEPA 3510C, 8270 C	Benzo(g,h,i)perylene	ug/l	5	<5	<5	<5	<5	<5	<5
USEPA 3510C, 8270 C	Phthalate Esters								
USEPA 3510C, 8270 C	Dimethyl phthalate	ug/l	5	<5	<5	<5	<5	<5	<5
USEPA 3510C, 8270 C	Diethyl phthalate	ug/l	5	<5	<5	<5	<5	<5	<5
USEPA 3510C, 8270 C	Di-n-butyl phthalate	ug/l	5	<5	<5	<5	<5	<5	<5
USEPA 3510C, 8270 C	Butyl benzyl phthalate	ug/l	5	<5	<5	<5	<5	<5	<5
USEPA 3510C, 8270 C	Bis(2-ethylhexyl) phthalate	ug/l	50	<50	<50	<50	<50	<50	<50
USEPA 3510C, 8270 C	Di-n-octyl phthalate	ug/l	5	<5	<5	<5	<5	<5	<5

LOR: Level of Reporting

Table A6.6: Continued 1

Method Reference	Analysis Description	Lab I.D		8242	8243	8244	8245	8246	8247
		Sample I.D		BH4/G	BH5/G	BH6/G	BH7/G	BH8/G	BH9/G
		Units	LOR	W	W	W	W	W	W
		Date of Extraction		5/4/2005					
		Date of Analysis		5/4/2005					
USEPA 3510C, 8270 C	Semivolatile Organic Carbon (SVOC)								
USEPA 3510C, 8270 C	Phenols								
USEPA 3510C, 8270 C	Phenol	ug/l	5	<5	<5	<5	<5	<5	<5
USEPA 3510C, 8270 C	2-chlorophenol	ug/l	5	<5	<5	<5	<5	<5	<5
USEPA 3510C, 8270 C	2-methylphenol	ug/l	5	<5	<5	<5	<5	<5	<5
USEPA 3510C, 8270 C	4-methylphenol	ug/l	5	<5	<5	<5	<5	<5	<5
USEPA 3510C, 8270 C	2-nitrophenol	ug/l	5	<5	<5	<5	<5	<5	<5
USEPA 3510C, 8270 C	2,4-dimethylphenol	ug/l	5	<5	<5	<5	<5	<5	<5
USEPA 3510C, 8270 C	2,4-dichlorophenol	ug/l	5	<5	<5	<5	<5	<5	<5
USEPA 3510C, 8270 C	2,6-dichlorophenol	ug/l	5	<5	<5	<5	<5	<5	<5
USEPA 3510C, 8270 C	4-chloro-3-methylphenol	ug/l	5	<5	<5	<5	<5	<5	<5
USEPA 3510C, 8270 C	2,4,6-trichlorophenol	ug/l	5	<5	<5	<5	<5	<5	<5
USEPA 3510C, 8270 C	2,4,5-trichlorophenol	ug/l	5	<5	<5	<5	<5	<5	<5
USEPA 3510C, 8270 C	Pentachlorophenol	ug/l	1	<1	<1	<1	<1	<1	<1
USEPA 3510C, 8270 C	Polynuclear Aromatics (PAH)								
USEPA 3510C, 8270 C	Naphthalene	ug/l	5	<5	<5	<5	<5	<5	<5
USEPA 3510C, 8270 C	Acenaphthylene	ug/l	5	<5	<5	<5	<5	<5	<5
USEPA 3510C, 8270 C	Acenaphthene	ug/l	5	<5	<5	<5	<5	<5	<5
USEPA 3510C, 8270 C	Fluorene	ug/l	5	<5	<5	<5	<5	<5	<5
USEPA 3510C, 8270 C	Phenanthrene	ug/l	5	<5	<5	<5	<5	<5	<5
USEPA 3510C, 8270 C	Anthracene	ug/l	5	<5	<5	<5	<5	<5	<5
USEPA 3510C, 8270 C	Fluoranthene	ug/l	5	<5	<5	<5	<5	<5	<5
USEPA 3510C, 8270 C	Pyrene	ug/l	5	<5	<5	<5	<5	<5	<5
USEPA 3510C, 8270 C	Benz(a)anthracene	ug/l	5	<5	<5	<5	<5	<5	<5
USEPA 3510C, 8270 C	Chrysene	ug/l	5	<5	<5	<5	<5	<5	<5
USEPA 3510C, 8270 C	Benzo(b) & (k)fluoranthene	ug/l	10	<10	<10	<10	<10	<10	<10
USEPA 3510C, 8270 C	Benzo(a)pyrene	ug/l	5	<5	<5	<5	<5	<5	<5
USEPA 3510C, 8270 C	Indeno (1,2,3-cd)pyrene	ug/l	5	<5	<5	<5	<5	<5	<5
USEPA 3510C, 8270 C	Dibenz(a,h)anthracene	ug/l	5	<5	<5	<5	<5	<5	<5
USEPA 3510C, 8270 C	Benzo(g,h,i)perylene	ug/l	5	<5	<5	<5	<5	<5	<5
USEPA 3510C, 8270 C	Phthalate Esters								
USEPA 3510C, 8270 C	Dimethyl phthalate	ug/l	5	<5	<5	<5	<5	<5	<5
USEPA 3510C, 8270 C	Diethyl phthalate	ug/l	5	<5	<5	<5	<5	<5	<5
USEPA 3510C, 8270 C	Di-n-butyl phthalate	ug/l	5	<5	<5	<5	<5	<5	<5
USEPA 3510C, 8270 C	Butyl benzyl phthalate	ug/l	5	<5	<5	<5	<5	<5	<5
USEPA 3510C, 8270 C	Bis(2-ethylhexyl) phthalate	ug/l	50	<50	<50	<50	<50	<50	<50
USEPA 3510C, 8270 C	Di-n-octyl phthalate	ug/l	5	<5	<5	<5	<5	<5	<5

LOR: Level of Reporting

Table A6.6: Continued 2

		Lab I.D		8242	8243	8244	8245	8246	8247
		Sample I.D		BH4/G	BH5/G	BH6/G	BH7/G	BH8/G	BH9/G
		Units	LOR	W	W	W	W	W	W
USEPA 3510C, 8270 C	1,2,4-Trichlorobenzene	ug/l	5	<5	<5	<5	<5	<5	<5
USEPA 3510C, 8270 C	Hexachloropropylene	ug/l	5	<5	<5	<5	<5	<5	<5
USEPA 3510C, 8270 C	Hexachlorobutadiene	ug/l	5	<5	<5	<5	<5	<5	<5
USEPA 3510C, 8270 C	Hexachlorocyclopentadiene	ug/l	25	<25	<25	<25	<25	<25	<25
USEPA 3510C, 8270 C	Pentachlorobenzene	ug/l	5	<5	<5	<5	<5	<5	<5
USEPA 3510C, 8270 C	Hexachlorobenzene	ug/l	10	<10	<10	<10	<10	<10	<10
USEPA 3510C, 8270 C	**Anilines and Benzidines**								
USEPA 3510C, 8270 C	Aniline	ug/l	5	<5	<5	<5	<5	<5	<5
USEPA 3510C, 8270 C	4-Chloroaniline	ug/l	5	<5	<5	<5	<5	<5	<5
USEPA 3510C, 8270 C	2-Nitroaniline	ug/l	10	<10	<10	<10	<10	<10	<10
USEPA 3510C, 8270 C	3-Nitroaniline	ug/l	10	<10	<10	<10	<10	<10	<10
USEPA 3510C, 8270 C	Dibenzofuran	ug/l	5	<5	<5	<5	<5	<5	<5
USEPA 3510C, 8270 C	4-Nitroaniline	ug/l	5	<5	<5	<5	<5	<5	<5
USEPA 3510C, 8270 C	Carbazole	ug/l	5	<5	<5	<5	<5	<5	<5
USEPA 3510C, 8270 C	3,3-Dichlorobenzidine	ug/l	5	<5	<5	<5	<5	<5	<5
USEPA 3510C, 8270 C	**Surrogate**	Units	Limits						
USEPA 3510C, 8270 C	2-Fluorophenol	%	30-61	60	48	46	49	46	49
USEPA 3510C, 8270 C	Phenol-d5	%	23-58	55	37	50	48	49	48
USEPA 3510C, 8270 C	2,4,6-Tribromophenol	%	67-123	96	81	83	87	81	90
USEPA 3510C, 8270 C	Nitrobenzene-d5	%	61-123	97	97	87	93	85	83
USEPA 3510C, 8270 C	2-Fluorobiphenyl	%	63-119	96	96	84	91	85	91
USEPA 3510C, 8270 C	4-Terpenyl-d4	%	68-117	97	85	84	95	94	97

LOR: Level of Reporting

Table A6.7: Test Results of Duplicate Sample for Metals in Groundwater

COMPOUND	QC DUPLICATE RESULTS		
	Sample Conc	Check Sample Conc	RPD
	ug/l	ug/l	%
Arsenic	<LOR	<LOR	-
Cadmium	<LOR	<LOR	-
Chromium	35	35	0
Lead	<LOR	<LOR	-
Selenium	<LOR	<LOR	-
Silver	<LOR	<LOR	-
Barium	32	35	9
Mercury	<LOR	<LOR	-

Table A6.8: Test Results of VOC in Duplicate Sample of Soil

COMPOUND	QC DUPLICATE RESULTS		
	Sample Conc	Check Sample Conc	RPD
	mg/kg	mg/kg	%
VOLATILE ORGANIC CARBON			
Tetrachloroethylene	<LOR	<LOR	-
1,1,1,2-Tetrachloroethane	<LOR	<LOR	-
1,1,2,2-Tetrachloroethane	<LOR	<LOR	-
1,2,3-Trichloropropane	<LOR	<LOR	-
1,2-Dibromo-3-chloropropane	<LOR	<LOR	-
Hexachlorobutadiene	<LOR	<LOR	-
Chlorobenzene	<LOR	<LOR	-
Bromobenzene	<LOR	<LOR	-
2-Chlorotoluene	<LOR	<LOR	-
4-Chlorotoluene	<LOR	<LOR	-
1,3-Dichlorobenzene	<LOR	<LOR	-
1,4-Dichlorobenzene	<LOR	<LOR	-
1,2-Dichlorobenzene	<LOR	<LOR	-
1,2,4-Trichlorobenzene	<LOR	<LOR	-
1,2,3-Trichlorobenzene	<LOR	<LOR	-
Chloroform	<LOR	<LOR	-
Bromodichloromethane	<LOR	<LOR	-
Dibromochloromethane	<LOR	<LOR	-
Bromoform	<LOR	<LOR	-

APPENDIX SEVEN

LOG OF BOREHOLES

It is a standard practice to maintain a log for any experiment that is being carried out. During the construction of the wells a log was also maintained. The log used during the research and investigative is shown in Figure A7.1.

It consists of the following headings;

Header - indicate the projection identification number, client, borehole number and the author who maintaining the log.

Subsurface profile - maintain records on the condition of the soil below the ground and indicate the depth. Provide some description on the type of soil such as type, colour, size, moisture, odour and density.

Sample - specify the depth and sample taken. Indicate the date and time. Soil samples are taken during the construction of the well and groundwater samples are taken upon development of the well. Using the PID indicate the VOC concentration. Well Details - make a sketch and label the well construction. Indicate the depth of well, depth interval of screened section, filter pack, bentonite and cement grout. Specification and equipment details - specify the well size, screen interval, depth of well, drilled by, drill date and method used.

Notes - record any observation and abnormalities during the process.

Maintaining a log provides the researcher the ability to trace the activities and verifying data and observation in the future. It also helps in report writing and to reconstruct the work in the future. A total of the 13 wells were constructed and the logs are represented in the subsequent pages.

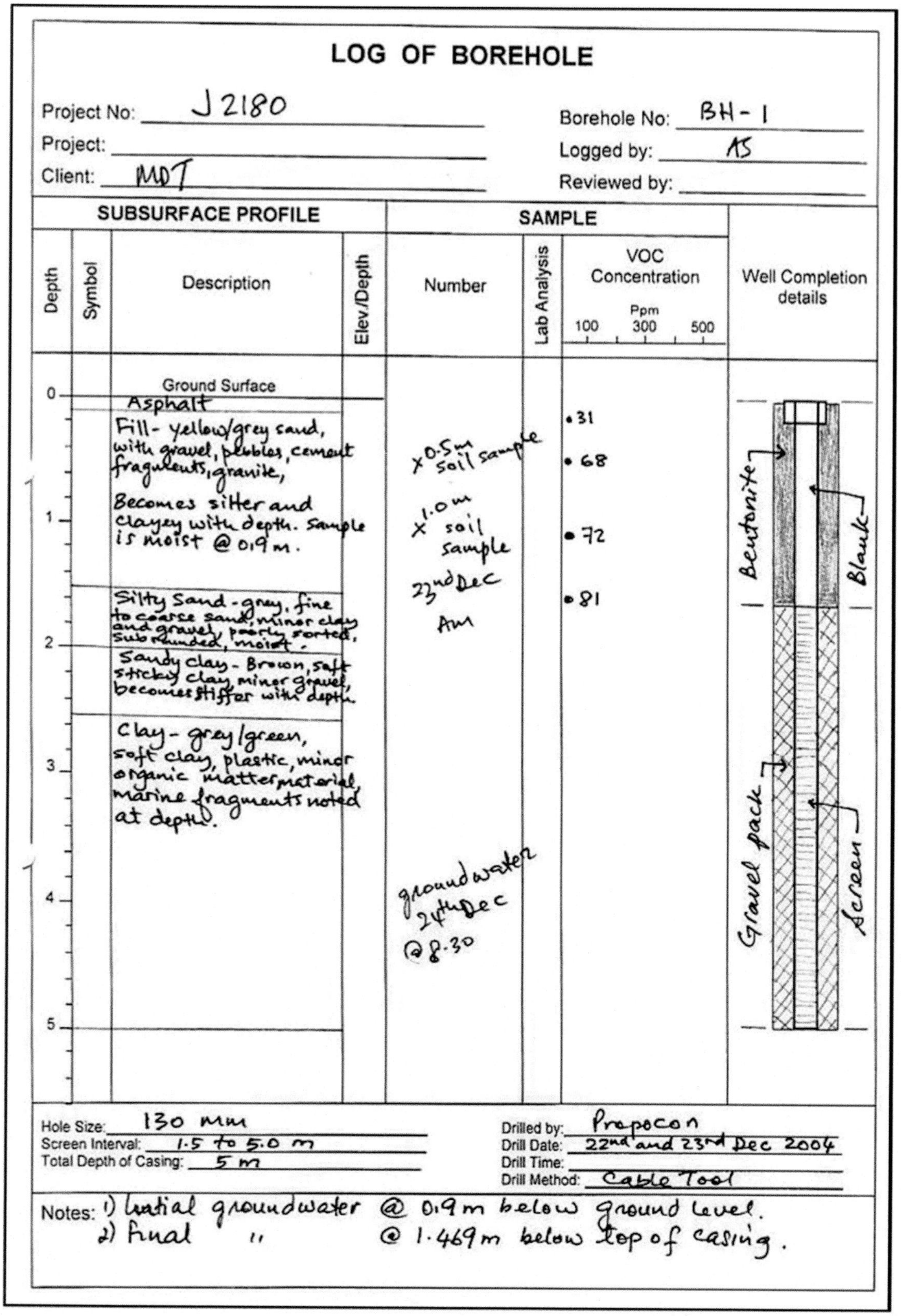

LOG OF BOREHOLE

Project No: J 2180

Borehole No: BH-1

Project: _____

Logged by: AS

Client: MDT

Reviewed by: _____

SUBSURFACE PROFILE				SAMPLE			Well Completion details
Depth	Symbol	Description	Elev/Depth	Number	Lab Analysis	VOC Concentration Ppm 100 300 500	
0		Ground Surface					
		Asphalt				•31	
		Fill- yellow/grey sand, with gravel, pebbles, cement fragments, granite,		X 0.5 m soil sample		•68	Bentonite / Blank
1		Becomes silter and clayey with depth. Sample is moist @ 0.19 m.		X 1.0 m soil sample		•72	
		Silty Sand - grey, fine to coarse sand, minor clay and gravel, poorly sorted, sub rounded, moist.		23rd Dec Am		•81	
2		Sandy clay - Brown, soft sticky clay, minor gravel, becomes stiffer with depth.					
3		Clay - grey/green, soft clay, plastic, minor organic matter/material, marine fragments noted at depth.					Gravel pack / Screen
4				groundwater 24th Dec @ 8.30			
5							

Hole Size: 130 mm
Screen Interval: 1.5 to 5.0 m
Total Depth of Casing: 5 m

Drilled by: Propocon
Drill Date: 22nd and 23rd Dec 2004
Drill Time: _____
Drill Method: Cable Tool

Notes: 1) Initial groundwater @ 0.19 m below ground level.
2) final " @ 1.469 m below top of casing.

Figure A7.1: Borehole Log BH1

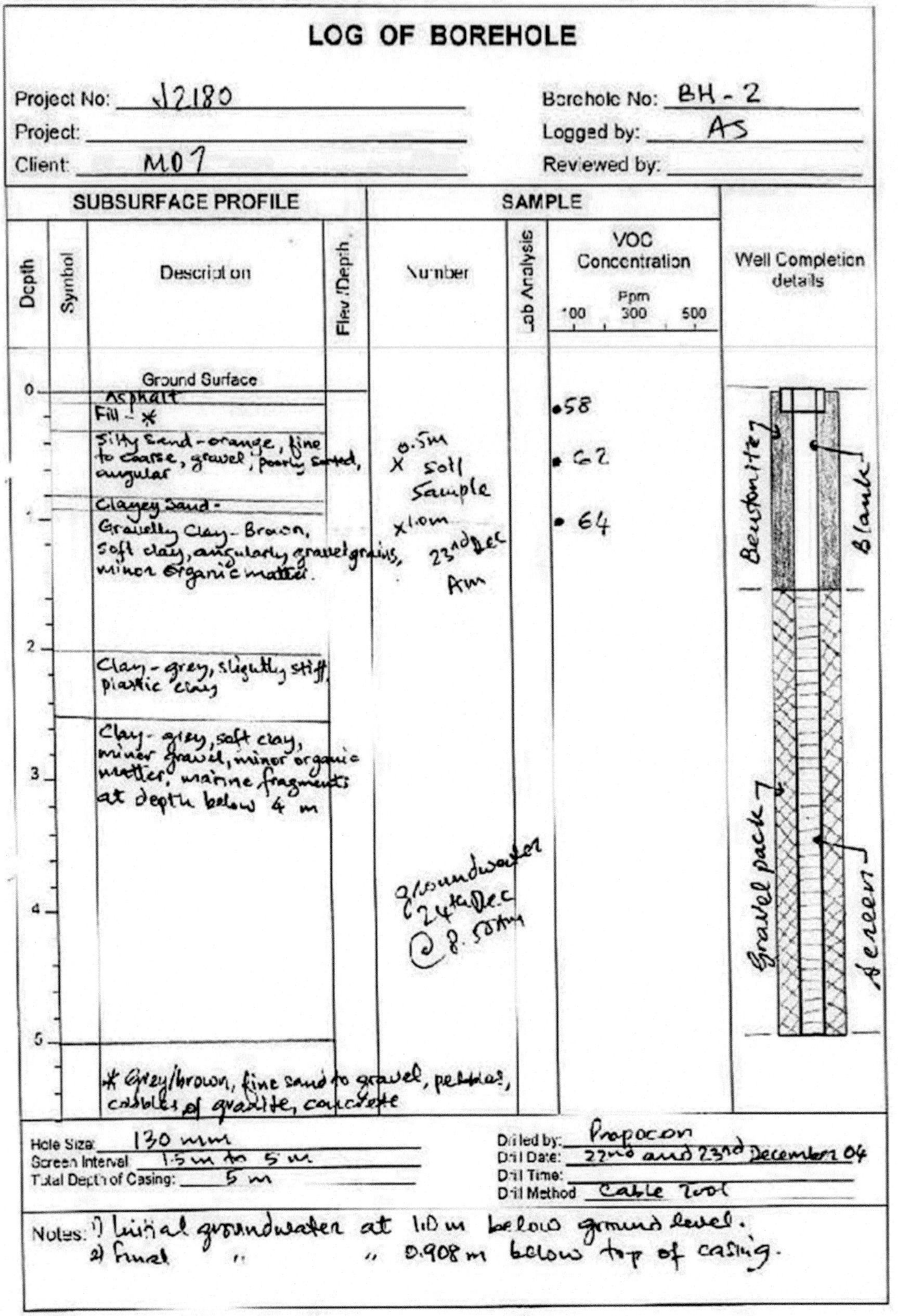

LOG OF BOREHOLE

Project No: √2180

Project: _____

Client: M07

Borehole No: BH-2

Logged by: AS

Reviewed by: _____

SUBSURFACE PROFILE | SAMPLE

Depth	Symbol	Description	Elev./Depth	Number	Lab Analysis	VOC Concentration ppm	Well Completion details

Ground Surface

Asphalt
Fill - *

Silty sand - orange, fine to coarse, gravel, poorly sorted, angular

Clayey Sand -
Gravelly Clay - Brown, soft clay, angularly gravel grains, minor organic matter.

Clay - grey, slightly stiff plastic clay

Clay - grey, soft clay, minor gravel, minor organic matter, marine fragments at depth below 4 m

0.5m
X soil
Sample

X 1.0m
23rd Dec
Am

groundwater
24th Dec
@ 8.50 Am

• 58
• 62
• 64

Bentonite

Blank

Gravel pack

Screen

* Grey/brown, fine sand to gravel, pebbles, cobbles of granite, concrete

Hole Size: 130 mm
Screen Interval: 1.5m to 5m
Total Depth of Casing: 5 m

Drilled by: Propocon
Drill Date: 22nd and 23rd December 04
Drill Time: _____
Drill Method: Cable tool

Notes: 1) initial groundwater at 1.0 m below ground level.
2) final " " 0.908 m below top of casing.

Figure A7.2: Borehole Log BH2

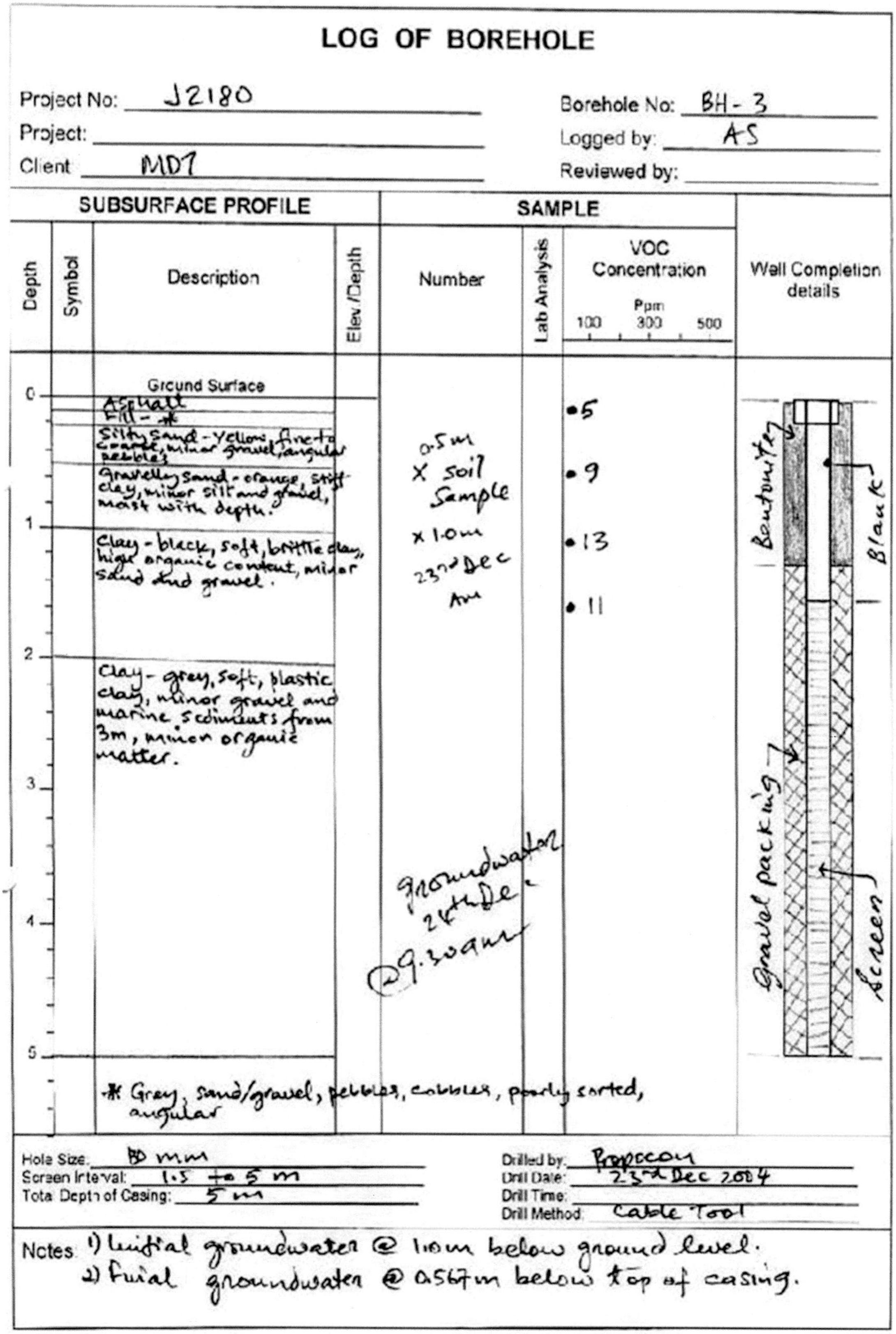

LOG OF BOREHOLE

Project No: _J2180_   Borehole No: _BH-3_

Project: _____   Logged by: _AS_

Client: _MD7_   Reviewed by: _____

SUBSURFACE PROFILE				SAMPLE			Well Completion details
Depth	Symbol	Description	Elev./Depth	Number	Lab Analysis	VOC Concentration Ppm 100 300 500	

Ground Surface

Description (by depth):
- Asphalt
- Fill - *
- Silty sand – Yellow, fine to coarse, minor gravel, angular pebbles
- Gravelly sand – orange, stiff clay, minor silt and gravel, moist with depth.
- Clay – black, soft, brittle clay, high organic content, minor sand and gravel.
- Clay – grey, soft, plastic clay, minor gravel and marine sediments from 3m, minor organic matter.

Sample Number:
- 0.5m X soil Sample
- X 1.0m 23rd Dec Am

VOC Concentration (Ppm):
- •5
- •9
- •13
- •11

groundwater 24th Dec @9.30am

Well Completion details: Bentonite, Blank, Gravel packing, Screen

* Grey, sand/gravel, pebbles, cobbles, poorly sorted, angular

Hole Size: _80 mm_   Drilled by: _Propecon_
Screen Interval: _1.5 to 5 m_   Drill Date: _23rd Dec 2004_
Total Depth of Casing: _5 m_   Drill Time: _____
                                       Drill Method: _Cable Tool_

Notes: 1) Initial groundwater @ 1.0m below ground level.
       2) Final groundwater @ 0.567m below top of casing.

Figure A7.3: Borehole Log BH3

LOG OF BOREHOLE

Project No: J2229

Project: _____

Client: MDT

Borehole No: BH-4

Logged by: MK

Reviewed by: AS

SUBSURFACE PROFILE

Depth	Symbol	Description	Elev./Depth
0		Ground Surface	
		Concrete	0.1
		Fill – grey, gravelly sand fine to med coarse grained sand, moist	0.5
		Fill – light brown, gravelly sand, fine to med coarse grain, poorly sorted, moist	1.0
1		Sandy clay – brown, streaks of red, soft, moderate plasticity, moist, no hydrocarbon odour	1.5
		Clay – grey, soft, high plasticity, wet @1.7m, no hydrocarbon odour.	
2			
3			
4			
5			

SAMPLE

Number	Lab Analysis	VOC Concentration Ppm 100 300 500	Well Completion details
		● 15.2	Bentonite 0.5
		● 13.2	Blank
		● 18.3	
		● 13.8	
1.7m X soil sample @ 10.50 3/31/05			Gravel pack Screen
groundwater @0915 3rd April			

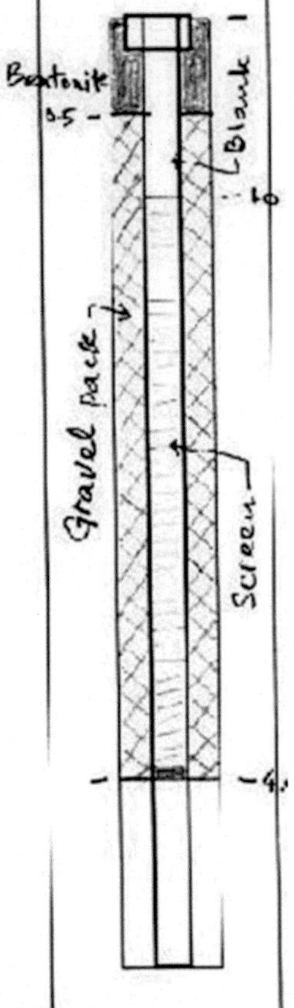

Hole Size: 100 mm
Screen Interval: 1-4 m
Total Depth of Casing: 4 m

Drilled by: Profocon
Drill Date: 31st March 2005
Drill Time: _____
Drill Method: Cable Tool

Notes:
1) Initial groundwater was 1.7 m below ground surface.
2) Final " " 1.210 m below top of casing

Figure A7.4: Borehole Log BH4

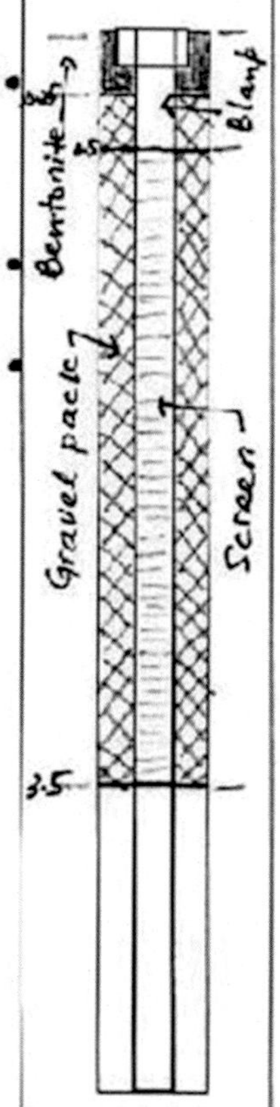

LOG OF BOREHOLE

Project No: J2229   Borehole No: BH-5
Project: _____   Logged by: AS
Client: MD1   Reviewed by: _____

SUBSURFACE PROFILE				SAMPLE			
Depth	Symbol	Description	Elev./Depth	Number	Lab Analysis	VOC Concentration Ppm 100 300 500	Well Completion details
0		Ground Surface					
		Asphalt	0.2			79,999 •	Blank
		Fill - brown, clayey gravel, coarse, moist, hydrocarbon odour, pebbles					Bentonite
1		Sandy Gravel - dark brown, coarse sand, poorly sorted, moist, hydrocarbon odour.		1.5 m X soil sample @ 11.13 3/31st		6173 • 898 •	Gravel pack
2		same as above					Screen
3		Sandy clay - Grey, soft, high plasticity, wet					
		Clay - Grey, soft, high plasticity, wet					3.5
4				Groundwater @ 0947 3rd April?			
5							

Hole Size: 100 mm
Screen Interval: 0.5 - 3.5 m
Total Depth of Casing: 3.5 m

Drilled by: Propocon
Drill Date: 31st March 2005
Drill Time:
Drill Method: Cable Tool

Notes: 1. Initial groundwater at 1.5 m below ground level.
2. Final groundwater at 0.987 m below top of casing.

Figure A7.5: Borehole Log BH5

LOG OF BOREHOLE

Project No: __J2229__   Borehole No: __BH-6__

Project: _____   Logged by: __AS__

Client: __MDT__   Reviewed by: _____

| SUBSURFACE PROFILE | | | | SAMPLE | | | | |
Depth	Symbol	Description	Elev./Depth	Number	Lab Analysis	VOC Concentration Ppm 100 300 500		Well Completion details
0		Ground Surface						
		Asphalt				79,999 •		
		Fill - *				9,044 •		
		Gravel sandy - reddish brown, moderate coarse, moderate sorted, subrounded, moist, slight hydrocarbon odour				79,999 •		
1		Sandy clay - light brown, friable, soft moist, hydrocarbon odour.		1.5 m X soil Sample @12.16 3/31/05		6.75 •		
2		Gravel sand - brown, stiff, moderate plasticity, subrounded to subangular, gravel, wet						
3		Clay. Grey, sticky, soft, high plasticity, wet						
4				groundwater @ 10.32 3rd April				
5		* Gravelly sand, fine to med coarse, subangular to subrounded, moist, hydrocarbon odour.						

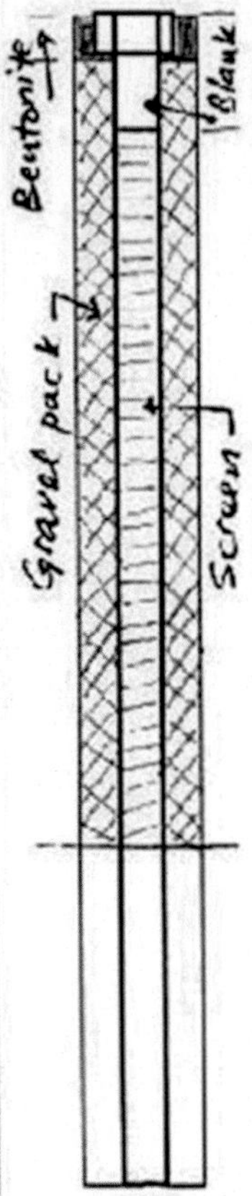

Hole Size: __100 mm__

Screen Interval: __0.5 to 3.5 m__

Total Depth of Casing: __3.5 m__

Drilled by: __Propocou__

Drill Date: __31st March 2005__

Drill Time: _____

Drill Method: __Cable tool__

Notes: 1) Initial groundwater at 4.5 m below ground level

2) Final groundwater at 0.666 m below top of casing.

Figure A7.6: Borehole Log BH6

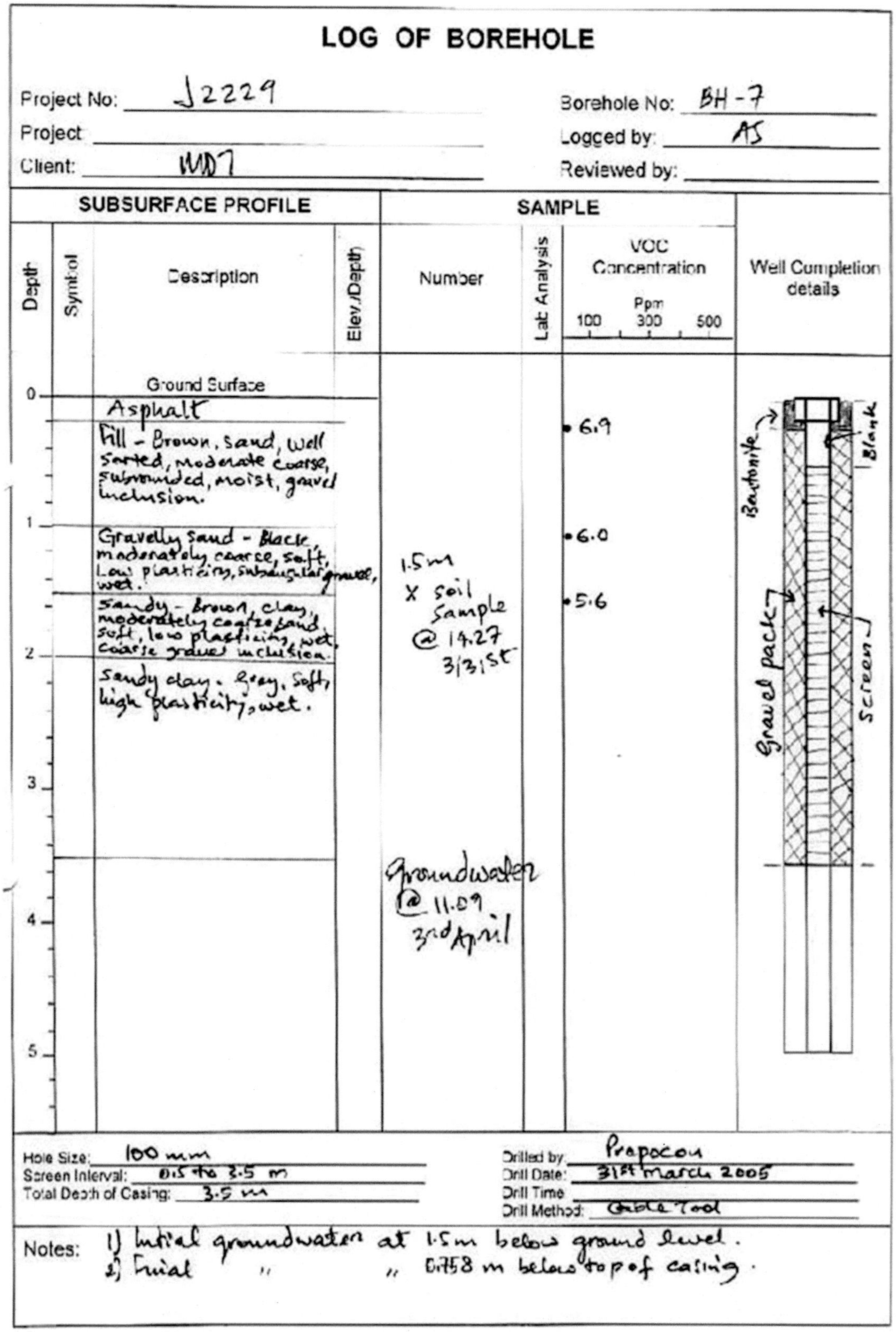

LOG OF BOREHOLE

Project No: ____J2229____   Borehole No: __BH-7__
Project _____   Logged by: ____AS____
Client: ____WD7____   Reviewed by: _____

SUBSURFACE PROFILE				SAMPLE			Well Completion details
Depth	Symbol	Description	Elev./Depth	Number	Lab Analysis	VOC Concentration Ppm 100 300 500	

Subsurface Profile descriptions:

0 — Ground Surface
Asphalt
Fill – Brown, sand, Well sorted, moderate coarse, subrounded, moist, gravel inclusion.
Gravelly sand – Black, moderately coarse, soft, Low plasticity, subangular gravel, wet.
Sandy – Brown, clay, moderately coarse sand, soft, low plasticity, wet, coarse gravel inclusion.
Sandy clay, grey, soft, high plasticity, wet.

Sample: 1.5m X soil Sample @ 14.27 3/3/5t

Groundwater @ 11.09 3rd April

VOC values:
• 6.9
• 6.0
• 5.6

Well Completion details labels: Bentonite, Blank, Gravel pack, Screen

Hole Size: ____100 mm____
Screen Interval: ____0.5 to 3.5 m____
Total Depth of Casing: ____3.5 m____

Drilled by: ____Prapocou____
Drill Date: ____31st March 2005____
Drill Time: _____
Drill Method: ____Cable Tool____

Notes:
1) Initial groundwater at 1.5m below ground level.
2) Final "   "   0.758 m below top of casing.

Figure A7.7: Borehole Log BH7

115

LOG OF BOREHOLE

Project No: J2229

Project: _____

Client: MM7

Borehole No: BH-8

Logged by: AS

Reviewed by: _____

		SUBSURFACE PROFILE			SAMPLE					Well Completion details
Depth	Symbol	Description	Elev./Depth		Number	Lab Analysis	VOC Concentration Ppm 100 300 500			

0		Ground Surface						
		Asphalt					• 4.7	
		Fill – Reddish brown, sand, fine to med coarse sand, moderately sorted, moist						
1		Gravelly clay – Grey, moderately coarse gravel, stiff, low plasticity, moist.					• 3.2	
					1.5m × soil sample @ 15.18 3/31st		• 4.9	
2		Gravelly clay – Yellowish brown w/ white streaks, moderately coarse gravel, soft, low plasticity, subangular gravel, wet						
3		Clay – Grey, soft, sticky, high plasticity, wet						
4					Groundwater @ 11.30 3rd April			
5								

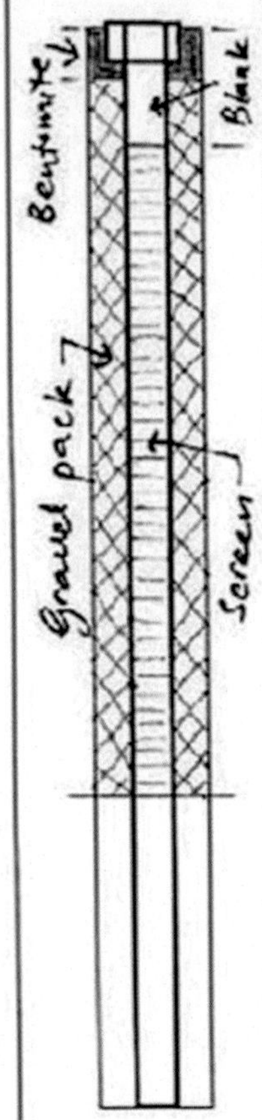

Hole Size: 100 mm
Screen Interval: 0.5 to 3.5 m
Total Depth of Casing: 3.5 m

Drilled by: Propocon
Drill Date: 31st March 2005
Drill Time: _____
Drill Method: Cable Tool

Notes: 1) Initial groundwater at 1.5m below ground level
2) Final   "   "   0.903m   "   top of casing.

Figure A7.8: Borehole Log BH8

116

LOG OF BOREHOLE

Project No. __J2229__   Borehole No: __BH-9__

Project _____   Logged by: __AS__

Client: __MCT__   Reviewed by: _____

		SUBSURFACE PROFILE		SAMPLE				
Depth	Symbol	Description	Elev./Depth	Number	Lab Analysis	VOC Concentration Ppm 100 300 500		Well Completion details
0		Ground Surface						
		Asphalt				• 2.8		
		Fill - *				• 3.3		
		Fill - light brown, sand, moderately sorted, subrounded, sand, slight moist no odour, gravel inclusion.				• 1.9		
1		Clayey Sand - Brown, fine sand, m. perately sorted, subrounded sand, soft clay, moist.		1.5m x soil sample @ 1648 3/3/5⁵		• 4.3		
2		Sandy clay - Grey, clay, soft, sticky, high plasticity, subrounded, wet, gravel inclusion						
3		Clay - Grey, soft, high plasticity, wet.						
4				groundwater @ 11.56 3rd April				
5		* Grey, gravelly sand, fine to medium grained, well sorted sand, subangular, to subrounded, moist.						

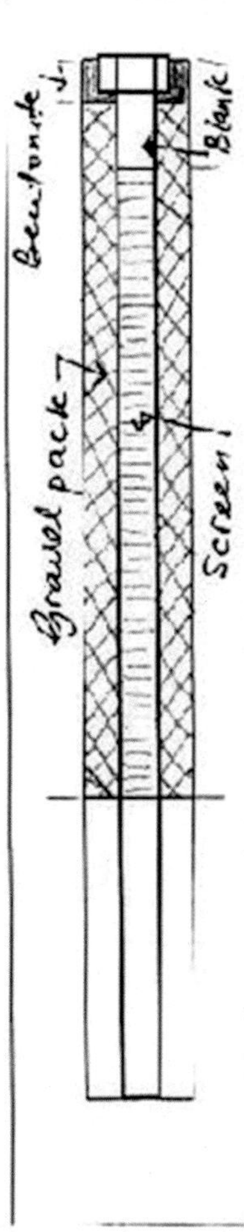

Hole Size: __100 mm__   Drilled by: __Propacon__

Screen Interval: __1.5 to 3.5 m__   Drill Date: __31st March 2005__

Total Depth of Casing: __3.5 m__   Drill Time: _____

Drill Method: __Cable Tool__

Notes: 1) Initial groundwater at 1.5 m below ground level.
2) Final   "   " 0.855 m below top of casing.

Figure A7.9: Borehole Log BH9

117

LOG OF BOREHOLE

Project No: __J 2229__   Borehole No: __BH-10__

Project: _____   Logged by: _____AS_____

Client: _____MD?_____   Reviewed by: _____

SUBSURFACE PROFILE			SAMPLE					
Depth	Symbol	Description	Elev./Depth	Number	Lab Analysis	VOC Concentration Ppm 100 300 500		Well Completion details
0		Ground Surface						
		Asphalt				• 4.2		
		Fill – brown, gravelly sand, med coarse, subangular, poorly sorted, moist, no odour.				• 2.8		
		Fill – brown, gravelly sand, fine to med coarse, moderately sorted, subangular/subrounded, moist				• 3.6		
1		Fill – Brown, gravelly sand, med coarse sand, moderately sorted, subangular gravel, moist				• 15.8		Bentonite
2		Clayey Sand – Moderately coarse sand, sticky clay, moderately sorted, wet, gravel inclusion.		1.6m X soil Sample @ 08:14 4/1st				Gravel pack
3		Clay – Dark gray, soft, high plasticity, wet						Screen
4				groundwater @ 12.20 3rd April				Blank
5								

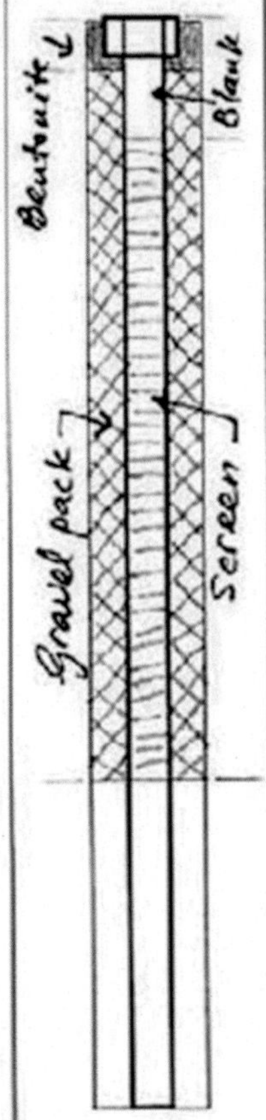

Hole Size: __100 mm__   Drilled by: __Propecon__

Screen Interval: __0.5 to 3.5 m__   Drill Date: __1st April 2005__

Total Depth of Casing: __3.5 m__   Drill Time: _____

                                     Drill Method: __Cable Tool__

Notes: 1) Initial groundwater at 1.6m below ground level.
2) Final "   "   0.434m " top of casing.

Figure A7.10: Borehole Log BH10

LOG OF BOREHOLE

Project No: __J2229__

Project: _____

Client: __MDT__

Borehole No: __BH-11__

Logged by: __AS__

Reviewed by: _____

	SUBSURFACE PROFILE			SAMPLE			Well Completion details
Depth	Symbol	Description	Elev /Depth	Number	Lab Analysis	VOC Concentration Ppm 100 300 500	

Subsurface profile descriptions (by depth):

- **0** Ground Surface
- Asphalt
- * Fill
- Asphalt
- Fill - gravel, subangular
- Clay - red, stiff, low plasticity, slightly moist
- **1** Gravelly clay - Reddish brown, moderately soft, moderately plasticity, moist.
- Clayey Sand - Brown, coarse, sticky clay, wet, gravel inclusion.
- **2**
- **3** Clay - Dark grey, soft plasticity, wet
- **4** * Fill - Gravel, subangular
- **5**

Sample notes:
- 1.5 m x soil sample @ 09.03 4/1st
- groundwater @ 12.43 3rd April

VOC Concentration:
- 12.1
- 5.1
- 2.1

Well Completion details: Bentonite, Blank, Gravel pack, Screen

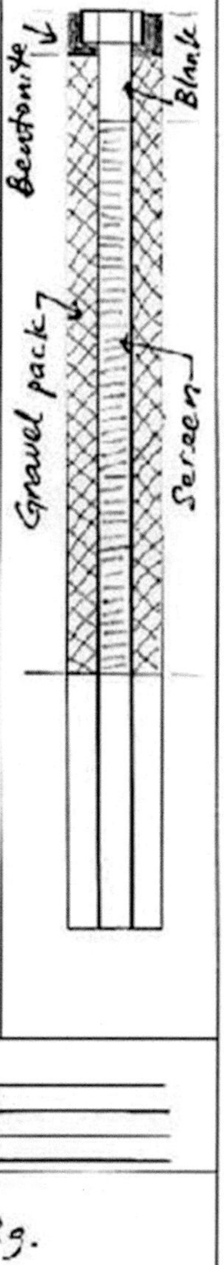

Hole Size: __100 mm__

Screen Interval: __0.5 to 3.5 m__

Total Depth of Casing: __3.5 m__

Drilled by: __Propocon__

Drill Date: __1st April 2005__

Drill Time: _____

Drill Method: __Cable Tool__

Notes: 1) Initial groundwater at 1.5 m below ground level
2) Final " " 0.642 m below top of casing.

Figure A7.11: Borehole Log BH11

119

LOG OF BOREHOLE

Project No: _J2229_   Borehole No: _BH-12_

Project: _____   Logged by: _AS_

Client _MDT_   Reviewed by: _____

SUBSURFACE PROFILE				SAMPLE				
Depth	Symbol	Description	Elev./Depth	Number	Lab Analysis	VOC Concentration (Ppm) 100 300 500		Well Completion details
0		Ground Surface Concrete						
1		Clayey Sand - brown, coarse, sticky clay, wet, gravel inclusion Sandy clay - * Clayey Sand - reddish brown, coarse, soft clay, low plasticity, subrounded, wet.		1.5m × soil sample @ 11.43 4/15t		• 1.2 • 0.3		Bentonite Blank
2		Gravelly Clay - # Clay - Dark grey, stiff, moderate plasticity, wet, clay is soft below 2.5m						Gravel pack Screen
3								
4				groundwater @ 14.10 3rd April				
5		* Brown, coarse, sticky, wet, gravel inclusion. # Brown, soft clay, low plasticity, subangular gravel.						

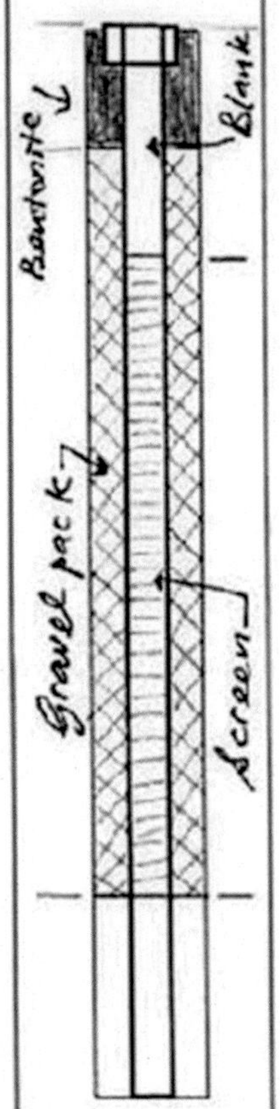

Hole Size _100 mm_

Screen Interval: _1.0 to 4.0 M_

Total Depth of Casing: _4.0 m_

Drilled by: _Propocon_

Drill Date: _1st April 2005_

Drill Time: _____

Drill Method: _Cable Tool_

Notes: 1) Initial groundwater 1.5m below ground level
2) final " " 1.174m " top of casing.

Figure A7.12: Borehole Log BH12

120

LOG OF BOREHOLE

Project No: __J2229__   Borehole No: __BH-13__

Project: _____   Logged by: __AS__

Client: __MDT__   Reviewed by: _____

SUBSURFACE PROFILE				SAMPLE			Well Completion details
Depth	Symbol	Description	Elev./Depth	Number	Lab Analysis	VOC Concentration Ppm 100 300 500	

Ground Surface

Concrete

Fill - gravelly sand

Fill - dark brown, clayey sand, fine to med coarse, low plasticity, slightly moist.

Gravelly Clay - reddish brown, soft, wet at 1.7m

Clay - Grey, moderately stiff, high plasticity, wet.

1.7m X soil Sample @ 1240 4/1st

groundwater @ 14.40 3rd April

● 1.9
● 2.0

● 2.9

Bentonite · Blank · Gravel pack · Screen

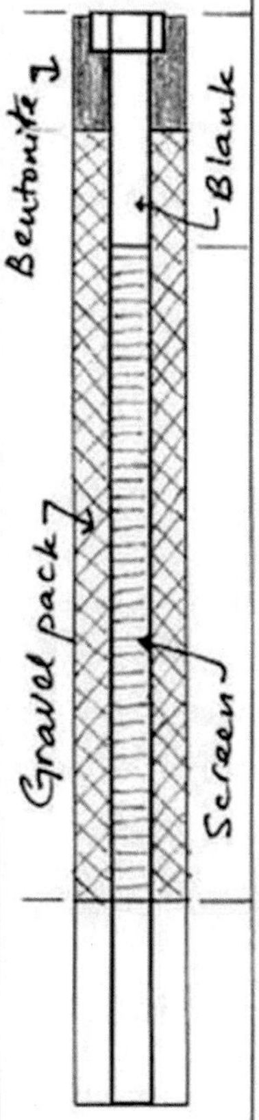

Hole Size: __100 mm__
Screen Interval: __1.0 - 4.0 m__
Total Depth of Casing: __4.0 m__

Drilled by: __Propocon__
Drill Date: __1st April 2005__
Drill Time: _____
Drill Method: __Cable Tool__

Notes: 1) Initial groundwater at 1.7m below ground level.
2) Final " at 1.321m below top of casing.

Figure A7.13: Borehole Log BH13

APPENDIX EIGHT

DUTCH GUIDELINES

Target Values

Target value of the standard is an indicative value at which there is a sustainable soil quality. This can be interpreted as the soil has to be at this value to completely recover its functional properties so that it will not harm the human, plant and animal life cycle. This is also an indicative benchmark for environmental quality in the long term that will have negligible risk to the ecosystem. These values were derived from extensive studies (HANS project) conducted in the Netherlands from 1996 to 1998 on unpolluted locations. These values were than adjusted so that most unpolluted locations will be able to meet this target value at a minimum of 95% of the time.

Target values for groundwater are published in Table 1 and 2 of the Dutch Standard. An extract of it is shown in Table A8.1. There is a distinction for metals in shallow or deep water. Deep water is considered to be greater than 10 meters. The standard also mentions that the 10-meter level is an arbitrary number and is only for indicative purposes. For shallow groundwater, the values have been adopted as target values. The background concentrations given in the tables are to be viewed as a guide. For some metals the background concentrations in shallow groundwater is higher than the surface water. Both the values are not coordinated, as such the authorizing authorities may have to decide whether the local target values need to be adjusted.

EOX Trigger Value

If the investigated site exceeds this value it does not mean that the site is polluted. It is required for the investigative team to conduct further investigations to determine the presence of the contaminants is not from natural causes. The Dutch Standards specify a protocol (NEN 5740) to conduct such investigations.

Table A8.1 shows the target values and intervention values for metal concentration in soil and groundwater.

	EARTH/SEDIMENT (mg/kg dry matter)			GROUNDWATER (µg/l in solution)			
	national background concentration (BC)	target value (incl. BC)	intervention value	target value shallow	national background concentration deep (BC)	target value deep (incl. BC)	intervention value
I Metals							
antimony	3	3	15	-	0.09	0.15	20
arsenic	29	29	55	10	7	7.2	60
barium	160	160	625	50	200	200	625
cadmium	0.8	0.8	12	0.4	0.06	0.06	6
chromium	100	100	380	1	2.4	2.5	30
cobalt	9	9	240	20	0.6	0.7	100
copper	36	36	190	15	1.3	1.3	75
mercury	0.3	0.3	10	0.05	-	0.01	0.3
lead	85	85	530	15	1.6	1.7	75
molybdenum	0.5	3	200	5	0.7	3.6	300
nickel	35	35	210	15	2.1	2.1	75
zinc	140	140	720	65	24	24	800

Table A8.1: (Table 1a of Dutch Standard). Target values and soil remediation intervention values and background concentrations in soil/sediment and groundwater for metals. Values for soil/sediment have been expressed as the concentration in a standard soil (10% organic matter and 25% clay).

Table A8.2 shows the target values and intervention values for inorganic compounds, aromatic compounds and PAH for soil and groundwater. This is not the complete list of the table. It is shown as a reference in the appendix for the reader to visualize what to expect when researching for more detailed information.

Table A8.2: (Table 1b of Dutch standard).

	EARTH/SEDIMENT (mg/kg dry matter)		GROUNDWATER (µg/l in solution)	
	target value	intervention value	target value	intervention value
II Inorganic compounds				
cyanides-free	1	20	5	1500
cyanides-complex (pH<5)[1]	5	650	10	1500
cyanides-complex (pH ≥5)	5	50	10	1500
thiocyanates (sum)	1	20	-	1500
bromide (mg Br/l)	20	-	0.3 mg/l*	-
chloride (mg Cl/l)	-	-	100 mg/l*	-
fluoride (mg F/l)	500[3]	-	0.5 mg/l*	-
III Aromatic compounds				
benzene	0.01	1	0.2	30
ethyl benzene	0.03	50	4	150
toluene	0.01	130	7	1000
xylenes	0.1	25	0.2	70
styrene (vinyl benzene)	0.3	100	6	300
phenol	0.05	40	0.2	2000
cresols (sum)	0.05	5	0.2	200
catechol(o-dihydroxybenzene)	0.05	20	0.2	1250
resorcinol(m-dihydroxybenzene)	0.05	10	0.2	600
hydroquinone(p-dihydroxybenzene)	0.05	10	0.2	800
IV Polycyclic aromatic hydrocarbons (PAH)				
PAH (sum 10) [4,14]	1	40	-	-
naphthalene			0.01	70
anthracene			0.0007*	5
phenatrene			0.003*	5
fluoranthene			0.003	1
benzo(a)anthracene			0.0001*	0.5
chrysene			0.003*	0.2
benzo(a)pyrene			0.0005*	0.05
benzo(ghi)perylene			0.0003	0.05
benzo(k)fluoranthene			0.0004*	0.05
indeno(1,2,3-cd)pyrene			0.0004*	0.05

Target values and intervention values for soil remediation soil/sediment and groundwater for inorganic compounds, aromatic compounds, PAH, chlorinated hydrocarbons, pesticides and other contaminants. Values for soil/sediment have been expressed as the concentration in a standard soil (10% organic matter and 25% clay). *"This is only part of the table of the Dutch Standard"*.

Table A8.3 shows the target values for metal concentration in soil and groundwater indicating serious contamination existing on site.

	EARTH/SEDIMENT (mg/kg dry matter)			GROUNDWATER (µg/l in solution)			
	national background concentration (BC)	target values (incl. BC)	indicative level serious contaminat-ion	target values shallow	national background concentratio n deep (BC)	target values deep (incl. BC)	indicative level serious contaminat-ion
I Metals							
beryllium	1.1	1.1	30	-	0.05*	0.05*	15
selenium	0.7	0.7	100	-	0.02	0.07	160
tellurium	-	-	600	-	-	-	70
thallium	1	1	15	-	<2*	2*	7
tin	19	-	900	-	<2*	2.2*	50
vanadium	42	42	250	-	1.2	1.2	70
silver	-	-	15	-	-	-	40

Table A8.3: Target values, indicative levels for serious contamination and background concentrations soil/sediment and groundwater for metals. Values for soil/sediment have been expressed as the concentration in a standard soil (10% organic matter and 25% clay).

APPENDIX NINE

PROJECT COSTING

The approach taken was to develop an investigation that would provide enough information that could be obtained to make the decision to buy or not to buy the property being investigated and also determine the liability level.

Several contractors were reviewed who could perform the drilling of wells. The short listed company was selected due to its experience in performing similar size jobs. They also had trained staff and produced reference from other jobs that they had completed. Several laboratories were short listed and reviewed for their capabilities. The short listed laboratory was selected due to experience in handling similar projects and they were able to conduct the analysis of samples per the specified protocols. Also, the Laboratory has subsidiaries established throughout the world. Cross reference with sister plants had provided positive reference for the laboratory. They were able to provide faster turnaround time and the service level was exceptionally good.

Table A9.1: Table Showing the Project Cost

Project Costing	Units	Cost/Unit RM	Total Cost RM
Well Installation	3	1600	4800
Permanent Well construction	3	400	1200
Mobilisation	1	1725	1725
Underground Service Locating			2000
Testing of Soil			
Metals	4	300	1200
VOC	4	600	2400
SVOC	4	600	2400
Testing of Groundwater			
Metals	4	300	1200
VOC	4	600	2400
SVOC	4	600	2400
Waste Collection & Disposal			
Expenses - containers for Soil and Water		500	500
Disposal to KA		2000	2000
Well Installation	10	1600	16000
Permanent Well construction	10	400	4000
Mobilisation	1	1725	1725
Underground Service Locating			4850
Testing of Soil			
Metals	11	300	3300
VOC	11	600	6600
SVOC	11	600	6600
Testing of Groundwater			
Metals	11	300	3300
VOC	11	600	6600
SVOC	11	600	6600
Waste Collection & Disposal			
Expenses - containers for Soil and Water		1500	1500
Disposal to KA		6000	6000
Testing - Lead in Paint	8	120	960
Asbestos Assessment	6	120	720
Total			**92980**

APPENDIX TEN

RISK ASSESSMENT

Exposure Routes - Conceptual Model

A qualitative approach for soil or groundwater impact prediction involves the consideration of fundamental sources of contamination occurrences, the pathway involved in the transfer of contamination and the impact on the receptor. Thus, exposure routes of contaminations are assessed based on source, pathway and receptor. So, a source need to exist, there need to be a pathway and a receptor needs to be present. When one of the three conditions is absent, than there is potentially no fear of any risk to human or to any living thing. The contamination is not transferable.

A summary of the above is as follows;

- The existence of a primary source for the contamination,

- Release mechanism of the primary contamination,

- Causing a secondary source of contamination,

- Release mechanism of the secondary contamination,

- A pathway is needed for the contamination to move to a receptor,

- The receptor can be a human or any biotic stream, and

- The means of transfer can be by inhalation, dermal contact or through ingestion.

The steps are outlined in a table format. Each contaminant is analysed on the table. All steps are completed from left to right and are filled up. The path is than traced and flow pattern is constructed.

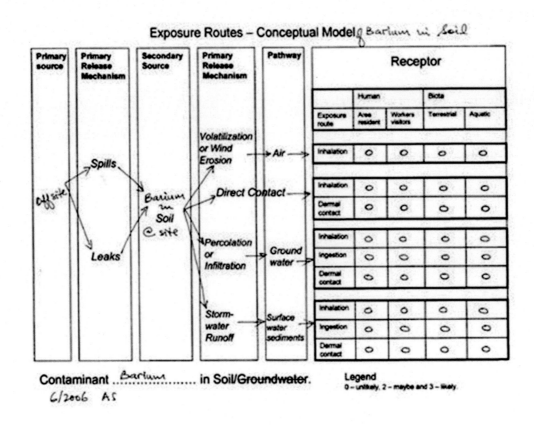

Figure A10.1: Exposure Routes - Conceptual Model for Barium in Soil

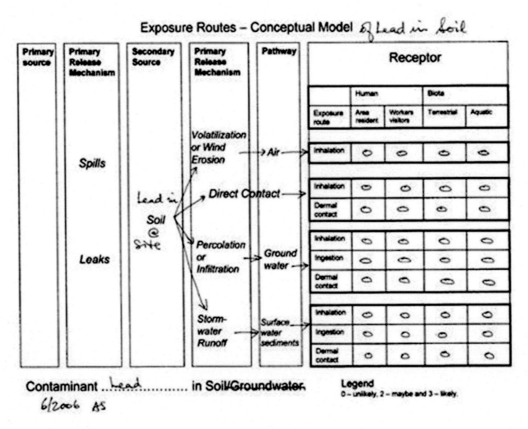

Figure A10.2: Exposure Routes - Conceptual Model for Lead in Soil

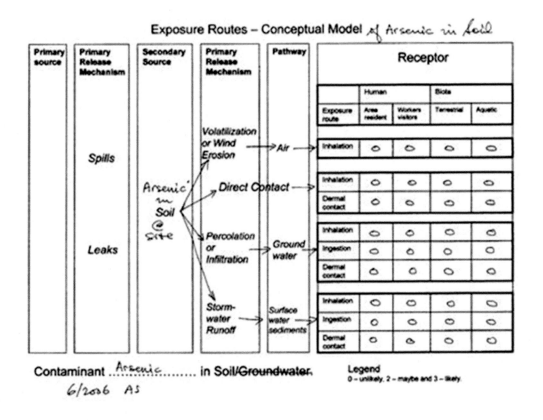

Figure A10.3: Exposure Routes - Conceptual Model for Arsenic in Soil

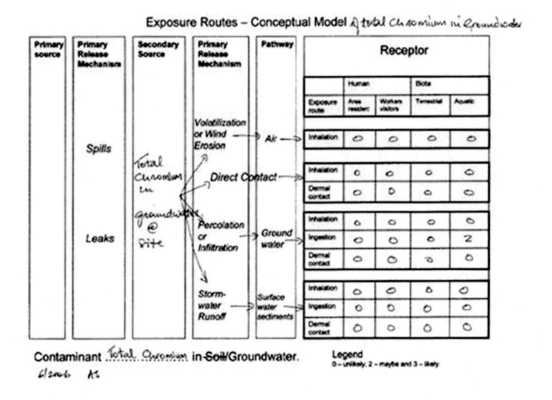

Figure A10.4: Exposure Routes - Conceptual Model for total Chromium in Groundwater.

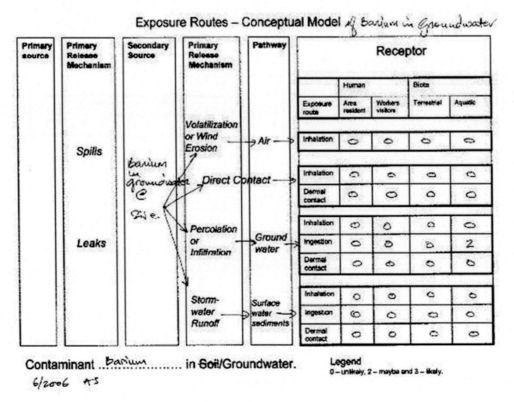

Figure A10.5: Exposure Routes - Conceptual Model of Barium in Groundwater.

Summary

The above contaminants do not have a pathway to the receptor. The contamination is 1 meter below ground level. The surface of the building is covered with concrete and the external surface is covered with bitumen. As such, the employees working on the site and visitors on the site will not be exposed to the contamination.

The concentration is limited to the south western side of the site.

The contamination being in the soil and groundwater does not pose any danger through inhalation. There are no drinking wells on site and the neighbouring sites. So, exposure through ingestion is not possible.

The employees and visitors on site will not be exposed as the surface of the site is covered with concrete and bitumen.

There is a possibility of exposure if construction work is conducted on site. If the need arises to conduct any construction work, the safety officer needs to draw up a protocol to wear appropriate personal protective equipment (PPE) for workers in the area To evaluate the receptor risk, each pathway is analysed with respect to human and biotic impact. A quantitative value is administered to determine the impact with in one another in the receptor boxes. The weighting is as follows;

- Value of '0' points for the event or impact to unlikely occur,

- Value of '2' points for the event or impact to 50/50 chance of occurring, and

- Value of '3' points for the event or impact to likely hood of occurring.

When the total numerical value is computed and analysed, for all the events or impacts to occur the value can go to a maximum of 108 points. Due to uncertainty or a probability of 50% chance of the event or impact to occur than the value will be 72 points. The severity of the event or impact will indicate a high point value.

To analyse each component and prioritizing the area of focus the figure below will assist in the decision making process. The analysis can be done cumulatively or per exposure route.

- Quadrant 1 – events or impacts that have low points and are low risk.

- Quadrant 2 – events or impacts that have low points and are at high risk.

- Quadrant 3 – events or impacts that have high points and are at low risk.

- Quadrant 4 – events or impacts that have high points and are at high risk.

The decision process will depend on the issues involved, the risk factor and cost implication.

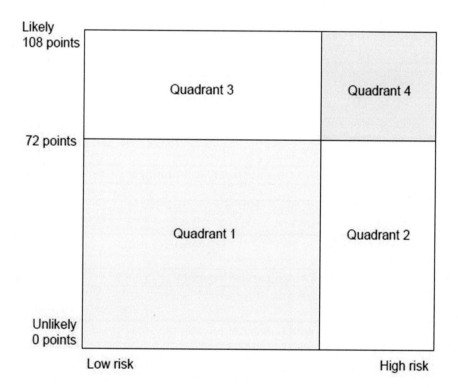

Figure A10.6: Risk Assessment of Exposure Routes for Conceptual Model.

APPENDIX ELEVEN

SUMMARY OF INTERVIEW RESULTS

Interview questionnaire

Name:

Years in Service:

Position: Store hand, HR Administrator, Finance Executive, Clerk.

Department:

Table A11.1: Part A: Investigated Site (Results in italics).

1.	Which location or process are you working? *store; administration; executive (Mr A); clerk;*
2.	How long have you been working at this process? *Years – 4; 6; 5; 2;*
3.	What is the name of your supervisor? *Mr A; Mr A; Mr Y; Ms A;*
4.	How many subordinates work for you? *Now none; now none, before 4; now 6, before 150; none;*
5.	What are the materials and chemicals used at this process? *Don't know; IPA, pcb, electronic components, solder, flux, wires; IPA, pcb boards, capacitors, transistors, diodes, transformers, IC, solder, flux, coils, pvc wires, copper wires; don't know;*
6.	Are you or your subordinate required to use Personal Protective Equipment (PPE) at your location/place of work? *All said no;*
7.	Which is/are the processes that you have seen using PPEs? *Chemical store area gloves and respirator.*
8.	Where materials/chemicals are stored prior to you or your subordinates using them? *Store*
	a. Do you know of any material/chemical spill at the location? *All said no;*
	b. Has your subordinates ever reported to you of material/chemical spill? *All said no;*
	c. Is case of a spill; is there any waste containment at the storage area? *2 said don't know and two said yes there is containment area.*
9.	Where do you store the materials/chemicals on the production line? *On wooden pallet and spill skid;*
	a. Did you observe any material/chemical spills at the location? *All said no;*
	b. Has your subordinates ever reported to you of material/chemical spill? *All said no;*
	c. Is case of a spill; is there any waste containment at the storage area? *2 said don't know and 2 said on spill skid;*
10.	What do you or your subordinates do with surplus material/chemicals? *2 said don't know and two said return to store;*

Table A11.1: Continued (Results in italics).

11. What do you or your subordinates do with waste material/chemicals? *2 said don't know and two said return to store;*	
12. Are the containers labeled ... part number, name, date, hazardous code, or any other identification? *All said containers are labeled;*	
13. How often do you or your subordinate transfer waste material/chemicals to waste storage area? *2 said very seldom and 2 said they do not know;*	
a. Did you observe any material/chemical spills at the location? *All said they had not seen any spill.*	
b. Has your subordinates ever reported to you of material/chemical spill? *All said none of their subordinates reported of any spill.*	
c. Is case of a spill is there any waste containment at the storage area? *One said don't know and three said yes there is containment for waste spills at the storage area.*	
14. Have you or your subordinates seen or heard of any spills occurring at the site? *Four said they do not know of any spill and neither has their subordinates ever reported of any spill.*	
15. Have you or your subordinates seen or heard of any employee disposing/throwing material/chemicals/waste in the drain, septic tank, toilets, over the fence, or at the council dump yard? *All four reported they do not know and neither has their subordinates reported of any.*	
16. Do you know how waste disposed from the site? *All four reported that waste is deposited into 44 gal drums and stored at the waste storage area.*	
17. Do you know who is responsible for waste disposal at your site? *All four reported that the store supervisor was responsible for waste disposal.*	
18. Have you ever seen a Kualiti Alam (KA) waste truck at the site? *All four reported that they have seen the KA waste collecting truck at the site.*	
19. Do you know or your subordinates know of any waste generated on site? *All four reported that there is waste generated on site. Waste is stored in containers but were not able to identify type of waste.*	
20. Do you wish to share anything with me? *All four did not have any further information to share.*	

Table A11.2: Part B: Surrounding Site

1.	What is your neighbor on the north of your site making/producing? *Don't know.*
	a. Do you know of any chemicals your neighbor is using? *No*
	b. Have you seen them throwing/disposing any material/chemical in the drain, septic tank, or on their premises? *No*
	c. Have you smelled of any odor of chemicals around the boundary? *No*
	d. Do you know where your neighbor stores their waste material? *No*
	e. Indicate location on layout? *All interviews were able to identify facility north of the investigated site.*
2.	What is your neighbor on the east of your site making/producing? *Electronic components – don't know what.*
	a. Do you know of any chemicals your neighbor is using? *No*
	b. Have you seen them throwing/disposing any material/chemical in the drain, septic tank, or on their premises? *No*
	c. Have you smelled of any odor of chemicals around the boundary? *No*
	d. Do you know where your neighbor stores their waste material? *No*
	e. Indicate location on layout? *All interviews were able to identify facility east of the investigated site.*
3.	What is your neighbor on the south of your site making/producing? *No*
	a. Do you know of any chemicals your neighbor is using? *No*
	b. Have you seen them throwing/disposing any material/chemical in the drain, septic tank, or on their premises? *No*
	c. Have you smelled of any odor of chemicals around the boundary? *No*
	d. Do you know where your neighbor stores their waste material? *No*
	e. Indicate location on layout? *All interviews were able to identify facility south of the investigated site.*
4.	What is your neighbor on the west of your site making/producing? *Three said glass products and one said lens.*
	a. Do you know of any chemicals your neighbor is using? *No*
	b. Have you seen them throwing/disposing any material/chemical in the drain, septic tank, or on their premises? *No*
	c. Have you smelled of any odor of chemicals around the boundary? *No*
	d. Do you know where your neighbor stores their waste material? *All four said yes.*
	e. Indicate location on layout? *All four showed the location of waste storage close to the boundary fence on the west side of the investigated site. All four were able to identify the location west of the investigated site.*
5.	Do you wish to share anything with me? *No*

Printed in the United States
By Bookmasters

An environmental audit on an industrial premise employing a due diligence approach was conducted which includes interview, site historical review, soil and groundwater sampling and analysis. The due diligence audit was conducted to ensure that the premise is free from any environmental and regulatory noncompliance, since there is a potential property transaction. Based on the initial investigation comprehending onsite and offsite interviews of workers from neighbouring sites and local authorities, findings indicate that the site is free from any contaminant.

Soil and groundwater samplings using borehole soil investigation method and further analysis using Inductive Couple Plasma (ICP) spectrometer for determining heavy metals, showed lead in paint at one location at a value of 870 ppm used in the building. Barium and lead was also detected in soil. Barium was 1020 and 576 ppm at two locations and lead was 914 and 880 ppm soil analysis. Chromium and barium was also detected in groundwater. Chromium was 56 and 63 ppm and barium was 1420 and 3780 ppm for groundwater.

The levels of the identified contaminants were compared with the Dutch Intervention (DIV 2000) Standards because Malaysia does not have any heavy metal standards for soil and groundwater contamination on any industrial premise.

A risk assessment was made to determine the extent of the potential harm that could arise from the contaminants and the impacts on the occupants on site, the neighbouring sites and the ecosystem. Based on the audit protocol for the investigation of a contaminated site developed in this study indicated that the risk assessment for the site is safe to be used.

PARTRIDGE

ISBN 978-1-5437-5067-6

90000

9 781543 750676

The Kitten Girls

are Scaredy Cats

Written by: Mickie Fosina

Illustrations: Susan Berger